Mastering the Patch Clamp
A Practical Guide

Fabian Christoph Roth · Markus Numberger ·
Andreas Draguhn

Mastering the Patch Clamp Technique: A Practical Guide

Springer

Fabian Christoph Roth
Medical Faculty, IMB University of Oslo
Oslo, Norway

Markus Numberger
GF Health Content
Kandel, Germany

Andreas Draguhn
Medizinische Fakultät Heidelberg
University
Heidelberg, Germany

ISBN 978-3-662-70744-9 ISBN 978-3-662-70745-6 (eBook)
https://doi.org/10.1007/978-3-662-70745-6

This book is a translation of the original German edition "Patch-Clamp-Technik," 2nd edition, by Fabian C. Roth et al., published by Springer-Verlag GmbH, DE in 2023. The translation was done with the help of an artificial intelligence machine translation tool. A subsequent human revision was done primarily in terms of content, so that the book will read stylistically differently from a conventional translation. Springer Nature works continuously to further the development of tools for the production of books and on the related technologies to support the authors.

Editorial Contact: Sarah Koch

This Springer imprint is published by the registered company Springer-Verlag GmbH, DE, part of Springer Nature.
The registered company address is: Heidelberger Platz 3, 14197 Berlin, Germany

Foreword

"Patch pipettes will be more useful than initially thought." Fred Sigworth's prediction has indeed come true. Originally, patch pipettes were used to measure "elementary" currents through single ion channels in the membrane of frog muscle fibers. Today, electrophysiologists use patch pipettes to investigate membrane currents, potentials, and capacitances in preparations as diverse as erythrocytes, plant cells, and neurons. Extended applications of the patch clamp technique now allow these processes to be measured with multiple pipettes in natural tissues and even in animals during behavior. Combined with modern imaging techniques, a detailed understanding of signaling in complex networks such as the cerebral cortex is within reach. This book provides a competent overview of the wide range of possibilities for recording electrical signals in living tissue with high resolution and, so to speak, to "eavesdrop" on the "conversations" between cells.

Bert Sakmann

Preface

In 1991, Erwin Neher and Bert Sakmann were awarded the Nobel Prize in Physiology or Medicine, in recognition of their work on the "function of single ion channels in cells". Fifteen years earlier, they had developed a method for measuring currents through single ion channels in the membrane of living cells. A discovery that, according to the German-British neurophysiologist and Nobel laureate Bernard Katz, "is equivalent to the detection of atomic particles 50 years ago in terms of both aesthetic satisfaction and scientific significance for biology".

The patch clamp technique is now one of the most important neurophysiological working methods. Its application has provided significant insights into the function and properties of ion channels in basic biomedical research, and it has also become indispensable in applied pharmacological research. For these reasons, the method plays an important role in the theoretical and practical training of biology, biochemistry, medicine, and pharmacy students, and in the context of bachelor's, master's, and doctoral theses.

This book is aimed at advanced students of the aforementioned subjects as well as scientists who wish to gain a deeper understanding of the technique and/or acquire the skills required to perform it. It provides concrete answers to the most important questions that arise, for example: What exactly is meant by "patching"? Which specific applications does the patch clamp technique offer? How do you set up a patch clamp rig? How do you proceed practically in your patch clamp experiment?

The first edition of this book was published more than 25 years ago. Since then, the patch clamp technique has further evolved. Therefore, almost all illustrations are new, and most chapters had to be largely rewritten. Some new parts, like Chap. 6 on special applications or Chap. 7 on the documentation and processing of data, have been added. Also in this edition, we have tried to facilitate access to the specialized literature by citing particularly relevant classic papers and important recent publications at the end of each chapter.

Oslo
Kandel
Heidelberg
December 2024

Fabian C. Roth
Markus Numberger
Andreas Draguhn

Acknowledgments

In order to include as many and detailed practical tips as possible, parts of the manuscript were sent to friends and colleagues for review. We would like to express our gratitude to all those who provided assistance by undertaking a careful review of the manuscript and offering a plethora of suggestions, additions, and illustrations. Specifically, we would like to mention the following colleagues (in alphabetical order).

1st Edition: Dr. Thomas Berger (Freiburg), Dr. Gabi von Blankenfeld (Turku), Astrid Düerkop (Berlin), Dr. Claudia Eder (Berlin), Dr. Siegrun Gabriel† (Berlin), Prof. Dr. Helmut Haas (Düsseldorf), Patricia Hoffmann (Berlin), Dr. Frank Kirchhoff (Berlin), Kerstin Lehmann (Berlin), Carsten Ohlemeyer (Berlin), Prof. Dr. Peter Ruppersberg (Tübingen), Dr. Rudolf Schubert (Rostock), Sebastian Schuchmann (Berlin), Dr. Monika Stengl (Regensburg), Dr. Peter Stern (Frankfurt), Prof. Dr. Walter Stühmer (Göttingen) and Dr. Heinz Terlau (Göttingen).

2nd Edition: Isabella Boccuni (Heidelberg), Dr. Claus Bruehl (Heidelberg), Dr. Alexei Egorov (Heidelberg), Bernd Polder (Tamm), PD Dr. Andre Rupp (Heidelberg) and Carina Knudsen (Oslo).

We would also like to take this opportunity to express our sincere thanks to the staff of Springer-Verlag, Heidelberg, for their professional and patient support.

Our special thanks go to Prof. Dr. Erwin Neher for his critical review of the manuscript of the 1st edition and his valuable suggestions and improvements, Prof. Dr. Bert Sakmann, who kindly agreed to write a Foreword, and Prof. Uwe Heinemann† (Berlin), without whose benevolent support the 1st edition would not have been possible.

Fabian C. Roth
Markus Numberger
Andreas Draguhn

Contents

The History of the Patch Clamp Technique

1

Electrical processes at biological membranes are universally prevalent. They contribute to elementary functions of osmoregulation, metabolism, and exocytosis, and are of central importance for electrical signaling in neurons, sensory and muscle cells. In short: Without knowledge of electrophysiological mechanisms, numerous life processes (and their dysfunctions in diseases) cannot be understood.

The function of nerve and muscle cells requires, among other things, ionic currents that flow through ion channels. With every movement, sensation, thought, and heartbeat, ion channels open and close. They play an important role not only in electrically excitable cells, but are present in every cell membrane of all organisms, including the membranes of cell organelles. The patch clamp technique, which we will introduce in this book, can be used to study single ion channels, as well as currents across the entire cell membrane, in great detail. Since the beginnings in the 1970s, it has become a widely used standard method for measuring electrical processes in cells. In this book, we want to present its theoretical basis, its practical implementation, and its many applications in a compact and "user-friendly" way.

1.1 Bioelectricity

The insight that nerve and muscle cells work with the help of electrical signals is relatively old. The Italian physician and naturalist Luigi Galvani (1727–1798) already believed in the existence of an "animal electricity" in his book published in 1791 *De viribus electricitatis in motu muscularis commentarius*. Despite numerous experiments, he could never conclusively prove it. He imagined that the surface of the muscle was charged with "one" electricity, its interior with "the other" electricity, and the nerve entering the interior formed a conductive bridge between them. It was not until 1838 that the Italian Carlo Matteucci (1811–1868)

F. C. Roth et al., *Mastering the Patch Clamp Technique: A Practical Guide*, https://doi.org/10.1007/978-3-662-70745-6_1

was able to directly measure the electrical current of a muscle. He claimed that the surface of a muscle had a positive voltage, its interior a negative voltage.

Through the comprehensive investigations on animal electricity (between 1848 and 1884) by Emil du Bois-Reymond (1818–1896), director of the Institute for Physiology at Charité in Berlin, Germany, the scientific field of electrophysiology was finally established. Du Bois-Reymond measured the nerve and muscle current with zinc/zinc sulfate electrodes and showed that when a nerve or muscle is stimulated, a "negative fluctuation" of this current occurs, meaning "that any part of the muscle that is in excitation behaves negatively relative to a resting part" (according to Bernstein 1912). He also disproved the prevailing belief that nerves could only conduct in one direction. Two years later, the German physicist and physiologist Hermann von Helmholtz (1821–1894) determined the speed of stimulus conduction in the frog nerve to be 26–30 m/s. However, how cells generate such electrical phenomena remained unknown for a long time.

1.2 The Ion Theory

By the end of the nineteenth century, it was already postulated that cells have a conductive cytoplasm and a lipid cell membrane, which is hardly electrically conductive, but permeable to water and many low molecular weight substances. Therefore, the Viennese physiologist Ernst von Brücke (1819–1892) suggested that this membrane could contain channels or pores that allow water to pass through, but exclude larger dissolved molecules. Later, the English physiologist William Bayliss (1860–1924) pointed out that such water-filled channels could also conduct ions without them having to lose their hydration shell. Finally, the German physiologist Julius Bernstein (1839–1917) developed the hypothesis that the nonuniform distribution of ions across a selectively permeable cell membrane is responsible for the resting potential. He calculated its amplitude – assuming that potassium plays the main role – to be –68 mV. He further wrote: "The membrane potential decreases with stimulation" and explained this by an increase in the ion permeability of the membrane (Bernstein 1912). His almost visionary "membrane theory of bioelectric currents", which he summarized in his book *Elektrobiologie* in 1912, should not obscure the fact that until the 1930s "the leading axonologists were thoroughly skeptical both of the membrane theory in general and of the local circuit theory in particular.“ (Hodgkin 1976).

1.2.1 The Voltage Clamp

It was not until the end of the 1930s that the American biophysicists Kenneth S. Cole (1900–1984) and Howard J. Curtis (1906–1972) developed the voltage clamp technique and thus demonstrated that the membrane conductivity of a nerve cell increases when excited. Cole (1979) later recalled this classic experiment: "Hodgkin visited us in our laboratory when we had just recorded the change in conductivity on the oscilloscope. He was as excited as I've ever seen him before, jumping up and down as we explained it."

Their experiments seemed to confirm Bernstein's hypothesis; they clearly showed that ionic currents were responsible for the electrical signals of nerve cells – the action potentials – but they did not reveal which ions were involved and how these currents were generated. Theoretically, the ions could passively flow through pores in the membrane or be actively moved through the membrane by transporter molecules (carriers, pumps).

1.2.2 Hodgkin and Huxley

The two Englishmen Alan Hodgkin (1914–1998) and Andrew Huxley (1917–2012) assumed a carrier model when, in the mid-1930s, they began to study the origin of the action potential on the giant axon of the squid. After a war-related interruption, they published a series of papers between 1949 and 1952 in which they showed how an action potential actually arises. Using the voltage clamp technique developed by Cole and Curtis, they discovered that neuronal excitation is caused by specific currents of sodium and potassium ions through the cell membrane, and were able to separate these ionic currents from each other using a mathematical model.

Hodgkin and Huxley had conducted their experiments to test the carrier hypothesis, but as Hodgkin later recalled, they were disappointed to find that this model was obviously wrong:

> "We had started off to test a carrier hypothesis and believed that even if that hypothesis was not correct, we should nevertheless be able to deduce a mechanism from the massive amount of electrical data that we had collected. These hopes faded as the analysis progressed. We soon realized that the carrier model could not be made to fit certain results, for example the nearly linear instantaneous current voltage relationship, and that it had to be replaced by some kind of voltage-dependent gate." (Hodgkin 1976)

Assuming such voltage-dependent and ion-selective "gates", Hodgkin and Huxley finally developed a series of equations that could explain the height and time course of the action potential with astonishing accuracy. For this achievement, the two scientists received the 1963 Nobel Prize in Physiology or Medicine (Hodgkin 1963; Huxley 1963). Hodgkin and Huxley were able to predict the most important properties of the voltage-dependent gates with their revolutionary work; however, it took more than 20 years before these gates (or channels, as we call them today) could actually be demonstrated.

1.3 The Development of the Patch Clamp Technique

This was roughly the state of research when Bert Sakmann and Erwin Neher were doing their doctoral theses at the Max Planck Institute for Psychiatry in Munich. In 1969 and 1970, as both later recalled (Neher and Sakmann 1992), “two fascinating papers” by Ross Bean, Steven Hladky and Denis Haydon appeared. They had observed the first ion channels in artificial membranes.

1.3.1 The First Channels using the Black-Film Technique

Bean, Hladky and Haydon constructed a measuring apparatus with two chambers separated by a Teflon wall with a small hole in it. They filled the two compartments with salt solution and then spread a drop of phospholipid solution over the hole in the Teflon wall. The lipid forms a thin film over the hole, similar to a soap bubble, which is compressed into a lipid bilayer by the external forces (buoyancy of the oil, water pressure and electrostatic attraction of the two bodies of water). Since this membrane appears optically black, it is called a black film or also after the inventors Müller-Rudin membrane. The hole is thus closed by an artificial lipid bilayer, which forms a very high electrical resistance for ions. However, if certain proteins, such as the bacterial antibiotic gramicidin A, are added to the salt solution in one of the two chambers, the ionic conductivity of the membrane increases abruptly: The peptide gramicidin A is incorporated into the artificial membrane and spontaneously forms ion channels. These open and close in an all-or-none fashion, thus causing short-term, abrupt changes in the current through the membrane.

These experiments showed for the first time the behavior of single ion channels. However, it was an artificial model and not an observation of natural channels from nerve or muscle cells. The first measurements of such channels came from the laboratory of Bernard Katz (1911–2003), a native of Leipzig, Germany, who – being persecuted by the Nazis as a Jew – emigrated to England in 1935 and later worked at University College in London (Katz 1986).

1.3.2 Noise Analysis

Katz had participated in the experiments of Hodgkin and Huxley in the 1930s and later discovered the processes underlying synaptic transmission at the neuromuscular junction, for which he received the 1970 Nobel Prize in Physiology or Medicine (Katz 1970). At the beginning of the 1970s, he and the Mexican neuroscientist Ricardo Miledi (1927–2017) were able to measure the properties of ion channels in the membrane of excitable cells for the first time. The two researchers succeeded in using an indirect and rather difficult-to-understand method, known as noise analysis, to determine some of the properties of the nicotinic acetylcholine receptor at the neuromuscular synapse, the motor endplate.

By analyzing the electrical noise (by "noise" we mean the random fluctuations in the ionic current induced by acetylcholine) the researchers were able to determine that a single acetylcholine receptor conducts about 10 million ions per second. This was an important clue to the underlying mechanism. If the acetylcholine receptor functioned like a transporter, it would have to move an ion across the membrane in 0.1 μs, a speed that seemed much too high for such a transport. Therefore, Katz and Miledi concluded that the current measured after acetylcholine application could not be caused by an active transport process, but that the ionic current must flow passively through channels.

The work of Katz and Miledi showed that ion channels in biological membranes have very similar properties to the gramicidin channels that had been observed in artificial membranes. However, their experiments were based on the simultaneous measurement of the current through a large number of ion channels. Therefore, they were not able to directly observe the opening and closing of single channels. As a result, Katz and Miledi could only draw indirect conclusions about the properties of the channels, such as the average time that each channel stays open (open time), and the average size of the current through each individual channel (current amplitude or single channel conductance). The direct observation of the activity of single channels was first achieved by Erwin Neher (*1944) and Bert Sakmann (*1942), who were awarded the Nobel Prize in Physiology or Medicine for this work in 1991 (Neher 1991; Sakmann 1991).

1.3.3 The First Patch Clamp Experiments

Sakmann had worked at University College in London with Bernard Katz from 1970 to 1973. After that, he went to the Max Planck Institute for Biophysical Chemistry in Göttingen, where Erwin Neher in the department of Hans Kuhn had already begun to investigate single channels in artificial membranes. Neher and Sakmann decided to study the acetylcholine receptors of frog muscle cells in addition to their current projects. They wanted to detect the single channels directly on the biological preparation!

However, this was not as easy as it sounds. The main problem was that the current through a single ion channel had to be extremely small, as expected from Katz's and Miledi's calculations. Even the electrical background noise was about 100 times greater than the current that Neher and Sakmann wanted to measure with the methods then in use. This background noise is mainly caused by the countless channels and ion transporters on the entire surface of the cell. How small the currents through individual channels are can, perhaps, be best illustrated by looking at the different current amplitudes in physiological processes (Fig. 1.1).

Neher and Sakmann wanted to extract the tiny single-channel currents from the background noise by electrically isolating a very small section of the cell membrane, a patch, from its surroundings. They therefore decided to place a very thin glass tube as a measuring electrode on the cell surface. The idea was to place this

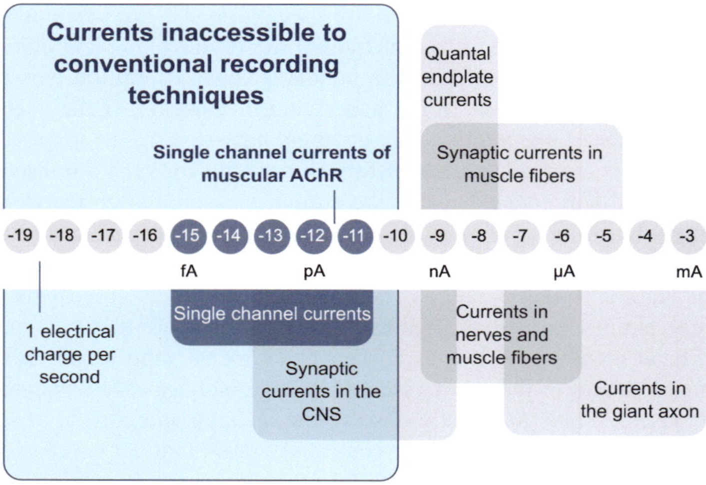

Fig. 1.1 Comparison of physiological and techical currents. The numbers are powers of ten of the unit Ampere. AChR: Acetylcholine Receptor, CNS: Central Nervous System

pipette over individual ion channels and electrically shield them from the environment, like under a cheese bell.

Such extracellular patch electrodes were first tested by Alfred Strickholm (1961) and later by Karl Frank in the early 1960s. Erwin Neher had already used similar "suction pipettes" in 1969 during his doctoral work in the laboratory of Dieter Lux in Munich. Neher recalled:

> "In Munich, the problem was not to measure the smallest possible currents absolutely, but to measure much larger currents with good relative resolution. In Göttingen, we then adapted the method to the new problem, namely good absolute resolution." (Neher, personal communication).

In order to achieve a good absolute resolution, the noise of the recording setup had to be minimized. This required a very tight connection between the cell membrane and the edge of the pipette to minimize "leak currents" between the pipette interior and the free extracellular space. The main problem was that the pipette could not get close enough to the cell surface because the basal lamina, an extracellular matrix surrounding the muscle fiber, and the connective tissue prevented a tight seal (Sigworth 1986).

Together with Sakmann, Neher was able to solve this problem. Sakmann had developed a method in Katz's lab in London to enzymatically free muscle cells from the basal membrane and the connective tissue components. Through this "cleaning" of the muscle cell surface, Neher and Sakmann were actually able to bring the measuring pipette into closer contact with the cell membrane, which increased the sealing resistance between the pipette and the external medium to about 10–50 MΩ (megaohms) and thus reduced the leak currents and the noise. In these experiments, the scientists used frog muscle fibers whose motor nerve had previously been severed. After denervation, the muscle cells form acetylcholine receptors not only locally in the area of the end plate, but on the entire muscle cell surface, albeit at a relatively low density – a favorable circumstance if one wants to measure only one or a few channels.

After many attempts, the seal was finally so good and the background noise so low that they could observe rectangular current pulses on the oscilloscope, which—as Neher put it in his Nobel lecture—"with good evidence, could be interpreted as single-channel currents" (Neher 1992; Fig. 1.2a). Neher continued:

> "The fact that similar records could be obtained both in our laboratory in Göttingen and in the laboratory of Charles F. Stevens at Yale (where I spent parts of 1975 and 1976), gave us confidence that they were not the result of some local demon, but rather signals of biological significance." (Neher 1992)

These rather inconspicuous, rectangular current pulses looked just like those already known from ion channels in artificial membranes (Neher and Sakmann 1976). For the first time, it was possible to directly observe the behavior of individual molecules in the membrane of a living cell! Erwin Neher later remembered:

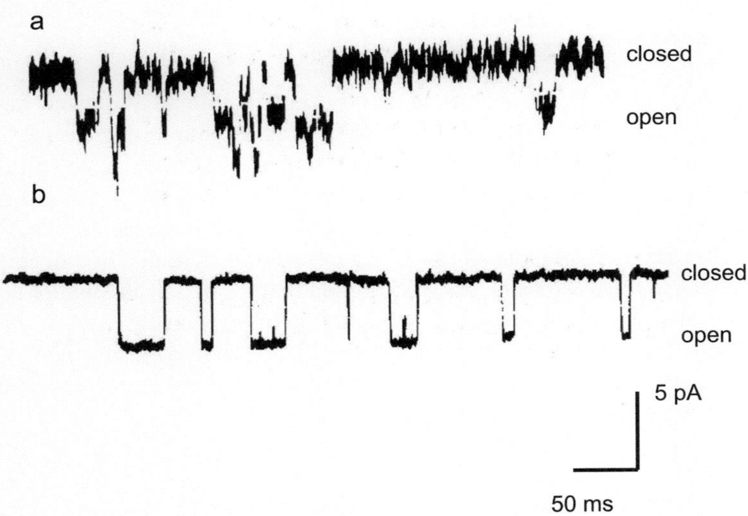

Fig. 1.2 Patch clamp measurements on single channels. **a** Recordings of nicotinic acetylcholine receptors made by Neher and Sakmann in 1975 during the first patch clamp experiments. **b** Single-channel openings of nicotinic acetylcholine receptors recorded by Neher and Sakmann in the early 1980s. Using the Gigaseal and improved amplifier technology, they were able to significantly reduce the background noise, which greatly improved the resolution of the single channel openings. (Image courtesy of Erwin Neher)

> "We started to optimize the method, i.e. to improve the contact between the pipette and the cell surface, mainly by different treatments of the cell surface or by changing the geometry of the pipette. Gradually, the noise became smaller, and we could see single channel openings. This was a continuous process that required many independent events to come together for it to work, and even then it was difficult to reproduce. At that time, we could only see the single channel currents under optimal conditions; as soon as we varied those conditions slightly, the currents disappeared. Nevertheless, by 1976 we had made some measurements on frog muscle, which Bert Sakmann first presented at a conference in Israel and which we then published in *Nature* in 1976." (Neher, personal communication)

The rectangular current pulses occur when a single acetylcholine receptor channel opens after binding of the transmitter and conducts ions for a certain time, i.e., when current flows. After a few milliseconds, the pore closes again, and the current abruptly returns to its original value. The measured currents appear on the oscilloscope as rectangular jumps (Fig. 1.2b), because the transitions between the open (ion-conducting) and the closed state occur extremely quickly. The measured currents thus show an all-or-none behavior: The channel is either completely closed or fully open and conducts a constant current in the open state. The amplitudes of the single channel events were therefore identical under the given conditions, only their duration varied. It was assumed that the ion channel opens as soon as acetylcholine binds to the receptor and closes when it is released. Today, we know that the mechanism must be described by somewhat more complicated kinetic models. The time during which the transmitter remains bound, and thus the open time of the ion channel, was apparently random and usually in the range of a few milliseconds.

The very first patch clamp recordings were already able to show the basic properties of ion channels. However, the noise was still quite strong in the first measurements made in the 1970s. Today, this technique of positioning the pipette very close to the membrane would be called loose-patch clamp (Sect. 6.1). In this way, only channels with a relatively high conductance can be measured. Another problem that apparently contradicted theoretical predictions was that not all single channel currents were of the same size. Occasionally, channels with smaller conductances appeared, which complicated the definition of the single channel current amplitude. Neher and Sakmann interpreted these as so-called edge channels, i.e., channels that were directly under the edge of the pipette and whose currents could therefore not be fully captured by the recording devices.

1.3.4 The Improvement of the Method: Gigaseals!

The solution to these problems, and thus the final breakthrough, was finally achieved by the two scientists after many efforts four years later. In January 1980, Neher serendipitously discovered that the sealing resistance between the pipette and the membrane increased by three orders of magnitude when he used a very clean, freshly pulled pipette and applied a slight negative pressure to the pipette upon contact with the cell surface. Unlike before, when the resistances between the bath and the pipette were at 10–50 MΩ, Neher suddenly obtained sealing resistances of up to 100 GΩ (gigaohms), which is why they are still called gigaseals. Neher did not work with denervated muscles at the time, but with cultured embryonic muscle cells, called myoblasts. He later described this as follows:

> "In the years between 1976 and 1980, we systematically tried to improve the sealing resistance between the pipette and the membrane in order to reduce the noise. We tested various enzymatic treatments of the cells, we tried to recharge the glass surface of the pipettes, to coat them with phospholipids or to make them hydrophobic in some other way – all more or less unsuccessfully. In between, we also repeatedly tried to suck, i.e., to apply negative pressure to the pipette. However, the improvement in sealing resistance was usually only slight, and the seal immediately collapsed again. So we finally gave up on these attempts and tried to get results with the existing technology.
>
> During these experiments, it occasionally happened that the pipette spontaneously formed a gigaseal upon contact with the cell surface. At that time, we did not know that this was possible, that a gigaseal could even exist, so we always thought that the pipette was 'clogged' and did not pay any further attention to the phenomenon. In November or December 1979, we had a visiting scientist to whom I wanted to show a patch experiment on myoblasts. Again, the pipette was "clogged", but suddenly we saw beautiful, clear channel openings on the oscilloscope. We immediately ran around the lab looking for TTX (tetrodotoxin, a sodium channel blocker) to see if they were really sodium channels, but by the time we found some, the cell was lost. Nevertheless, for the first time we had a recording with a gigaseal and then we looked for "clogged" pipettes, so that we could get such recordings from time to time. Unfortunately, this was not reproducible, and we had no idea why a gigaseal sometimes formed or how to induce it.
>
> Today we know that it was due to our pipettes. I have to say that until about 1978, we made our patch pipettes very laboriously. The pipette was first drawn normally, then it was clamped into a *microforge* (a device for post-treatment of micropipettes), and the tip was bent into a hook. A weight was hung on this hook and the pipette was pulled out again. The manufacturing process took about 10 to 15 min per pipette, so we used the pipettes several times, four or five times, until they were completely clogged with dirt and membrane fragments. Eventually, however, I realized that good quality pipettes can also be made using a conventional pipette puller and two pulling steps. This was much easier and faster, but we continued to use the pipettes several times in a row.
>
> On a Saturday in January 1980, I was working in the lab in the afternoon and had made a whole series of pipettes. I actually wanted to go home quickly, but I also wanted to use up the pipettes (they have to be used soon after they are made). So I used each pipette only once that day. That was the key change: In five tries, I suddenly got four gigaseals!
>
> In the following week, all our coworkers used fresh patch pipettes for each attempt and got gigaseals more and more often. This went well for a few weeks, but suddenly nothing worked at all for two weeks, and great frustration spread among all of us as no one knew what could be the cause. Eventually, we found out that a slight positive pressure had to be applied to the pipette to keep it from getting clogged with dirt in the bath solution. From then on, everything was fine." (Neher, personal communication)

This discovery was due not only to the new preparation – seals form more easily on cultured muscle cells – but mainly to the fact that Neher used a new, freshly pulled pipette for each recording. Once the gigaseals, which the Göttingen team were now receiving more frequently and more quickly, had almost completely eliminated the electrical noise of the biological preparation, other sources of noise became a limiting factor. Thus, it was necessary to reduce the noise caused by the electronic components of the amplifier and by the recording pipette itself. Neher says:

> "In the early experiments, before the discovery of the gigaseal, we did not make high demands on the amplifiers, we simply further developed the conventional current-voltage converters from the black-film technology. With the gigaseal, however, we definitely needed much better amplifiers. The main improvements were to increase the bandwidth and to optimize the noise at the high frequencies that were not accessible before. Fred Sigworth took care of the amplifier development and, after about four in-house prototypes, built the EPC-5, the first commercial patch clamp amplifier." (Neher, personal communication)

1.4 The Further Development of the Technique

The physics of a gigaseal is still not fully understood. Surprisingly, the membrane "sticks" very firmly to the mouth or the inner glass wall of the pipette. This allows various manipulations with a membrane patch without breaking the connection between the glass wall and the membrane (Fig. 1.3). Thus, not only can channel openings be measured by placing the pipette on the cell membrane (cell-attached configuration), but the piece of membrane in the pipette tip can also be broken through by a short suction pulse (whole-cell configuration). This gives open access to the cytoplasm, resulting in a configuration equivalent to the conventional intracellular recording, as both methods measure the currents across the entire cell membrane.

In addition, Bert Sakmann and Owen P. Hamill found that by retracting the pipette, a piece of membrane can be torn from the cell membrane. The channels present in this membrane patch can then be measured under cell-free conditions. By various manipulations, these pieces of membrane can be arranged so that either the originally cytoplasmic side of the cell membrane faces the bath solution (inside-out configuration) or the extracellular surface (outside-out configuration). Details of the different configurations are discussed in Chap. 5. The four most important configurations were published by the Göttingen scientists in 1981 in the seminal paper in *Pflügers Archiv* (Fig. 1.3). Neher comments:

> "Once we were able to obtain gigaseals reproducibly, we quickly developed the various configurations of the patch clamp technique from January to March 1980, which we then summarized in the publication for *Pflügers Archiv* in March 1981. I remember one of the reviewers complaining that we had included the whole-cell configuration. He thought it was off-topic and probably just an insignificant "frill". To accommodate him, we then removed an illustration (the one for the current clamp mode). Today I regret that, we should not have done it." (Neher, personal communication)

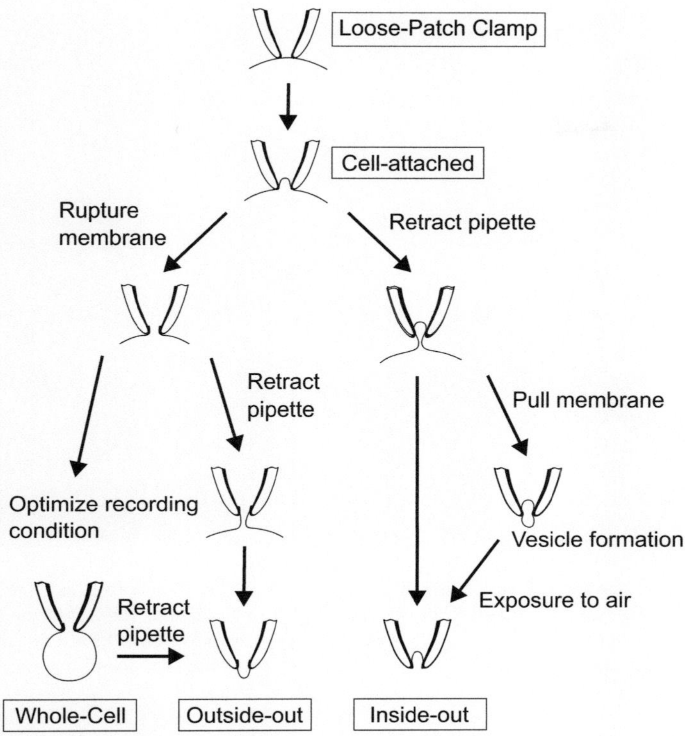

Fig. 1.3 The various patch configurations as published by Neher and Sakmann in their seminal work in *Pflügers Archive*. (After Hamill et al. 1981)

The group of Neher and Sakmann in Göttingen (Fig. 1.4) developed the patch clamp technique to study single ion channels in biological membranes. The technique proved so useful and universally applicable that it was quickly adopted by many scientists around the world and has been used to study many important ion channels and neurotransmitter receptors in a wide variety of cells. Since the introduction of the patch clamp technique, many other applications have emerged. It has become the most widely used method in cellular electrophysiology, not only to elucidate elementary processes at the molecular level, but also to address physiological questions at the level of neuronal networks (Sakmann 2017).

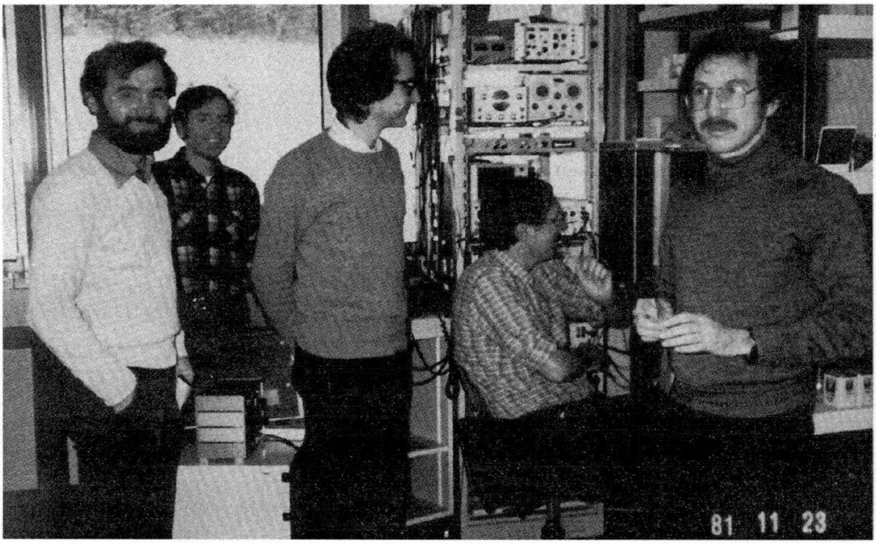

Fig. 1.4 The research group of Neher and Sakmann at the Max Planck Institute for Biophysical Chemistry in Göttingen, Germany. The photo was taken in November 1981 after the publication of the decisive paper in *Pflügers Archive*. From left to right: Erwin Neher, Fred J. Sigworth, Alain Marty, Bert Sakmann and Owen P. Hamill. (Photo courtesy of Bert Sakmann)

References

Bean RC, Shepherd WC, Chan H, Eichner J (1969) Discrete conductance fluctuations in lipid bilayer protein membranes. J Gen Physiol 53:741–757

Bernstein J (1912) Elektrobiologie. Vieweg, Braunschweig

Cole KS (1979) Mostly Membranes. Ann Rev Physiol 41:1–24

Hamill OP, Marty A, Neher E, Sakmann B, Sigworth FJ (1981) Improved patch-clamp techniques for high-resolution current recording from cells and cell-free membrane patches. Pflügers Arch 391:85–100

Hladky S, Haydon D (1970) Discreteness of conductance change in bimolecular lipid membranes in the presence of certain antibiotics. Nature 225:451–453

Hodgkin AL (1963) The ionic basis of nervous conduction. The Nobel Lecture. http://nobelprize.org/nobel_prizes/medicine/laureates/1963/hodgkin-lecture.pdf

Hodgkin AL (1976) Change and design in electrophysiology: an informal account of certain experiments on nerve carried out between 1934 and 1952. J Physiol 263:1–21

Huxley AF (1963) The quantitative analysis of excitation and conduction in nerve. The Nobel Lecture. http://nobelprize.org/nobel_prizes/medicine/laureates/1963/huxley-lecture.pdf

Katz B (1970) On the quantal nature of neuronal transmitter release. The Nobel Lecture. http://nobelprize.org/nobel_prizes/medicine/laureates/1970/katz-lecture.pdf

Katz B (1986) Reminiscences of a physiologist, 50 years after. J Physiol 370:1–12

Neher E (1992) Nobel Lecture—Ion channels for communication between and within cells. EMBO J 11:1673–1679

Neher E, Sakmann B (1976) Single-channel currents recorded from membrane of denervated frog muscle fibers. Nature 260:799–801

Neher E, Sakmann B (1992) The patch clamp technique. Sci Am 266:44–51

Sakmann, B (1991) Elementary steps in synaptic transmission revealed by currents through single ion channels. The Nobel Lecture. http://nobelprize.org/nobel_prizes/medicine/laureates/1991/sakmann-lecture.pdf

Sakmann B (2017) From single cells and single columns to cortical networks: dendritic excitability, coincidence detection and synaptic transmission in brain slices and brains. Exp Physiol 102:489–521

Sigworth FJ (1986) The patch clamp is more useful than anyone had expected. Fed Proc 45:2673–2677

Strickholm A (1961) Impedance of a small electrically isolated area of the muscle cell surface. J Gen Physiol 44:1073–1088

Further Reading

Bear MF, Connors BW, Paradiso MA (2020) Neuroscience. Exploring the brain. (4 edn) Jones and Bartlett Publishers, Inc

Hille B (2001) Ionic channels of excitable membranes. (3 edn) Sinauer, Sunderland

Kandel E, Schwartz JH, Jessell TM (2020) Principles of neural science. (5 edn) McGraw-Hill Education Ltd

Bioelectrical Phenomena and their Measurement

2

Electrophysiology plays a special role within the biosciences because its methods and concepts are very specific and difficult to understand for outsiders. Some people claim that you can recognize an electrophysiologist's lecture by the fact that most of the audience's eyelids drop. But: Anyone reading this book obviously does not want to be in that group. So don't worry—on the following pages we'll show you that electrophysiology is not esoteric. On the contrary—with simple logical thinking, Ohm's law and, above all, some practice, you can understand its techniques and diagrams very well. A degree in physics is definitely not required!

We will now introduce the basic electrical phenomena in cells, then describe the most important recording methods and illustrate them with examples.

2.1 The Basics

2.1.1 The Membrane Potential

All cells have a membrane potential that results from the unequal distribution of ions between the intra- and extracellular space. The membrane thus plays a prominent role in separating these two spaces. The properties of the membrane determine the entire exchange of substances and charges between the cell and its environment.

We derive the membrane potential as a potassium diffusion potential, because potassium ions have the greatest influence on the resting membrane potential in most cells. Our considerations can then easily be extended to other ions. So, let's imagine that a cell membrane has only one type of ion channel, and that it is exclusively permeable to potassium. Now we need a process that leads to an unequal distribution of potassium ions: the activity of the sodium–potassium pump (Na^+ /K^+-ATPase). This pump moves potassium ions (K^+) into the cell and sodium

F. C. Roth et al., *Mastering the Patch Clamp Technique: A Practical Guide*,
https://doi.org/10.1007/978-3-662-70745-6_2

ions (Na^+) out of the cell, using the energy of ATP hydrolysis. This results in a high concentration of K^+ of about 150 mM inside the cell, while in the extracellular space of mammals there is only about 3–5 mM K^+. (At this point, we briefly have to consider other ion species: this very unequal distribution of K^+ is, of course, only possible if it is largely compensated by other ions. In our case, this is mainly Na^+, which is much more concentrated outside than inside due to the Na^+/K^+-ATPase.) Because of the higher concentration of K^+ inside the cell, it is much more likely that a potassium ion will flow through open potassium channels from the inside to the outside than vice versa. As a result, the cell loses positively charged ions and becomes negatively charged on the inside. This negative membrane potential in turn leads to an electrostatic attractive force that pulls K^+ ions from the outside to the inside (the term "force" is figurative, physically correct would be to talk about currents). At a certain value of the potential, the diffusion flow to the outside and the electrically driven flow to the inside are in equilibrium—we speak of the "equilibrium potential" and calculate it using the Nernst equation:

$$E_{K+} = -RT/zF \times \ln\left(\left[K^+\right]_i / \left[K^+\right]_o\right)$$

E_{K+} is the potassium equilibrium potential, R is the universal gas constant, T is the absolute temperature in Kelvin, z is the valence of the ion (for $K^+ = 1$), F is the Faraday constant, $[K^+]_i$ is the K^+ concentration inside the cell and $[K^+]_o$ is the K^+ concentration outside the cell. For 37 °C and monovalent potassium ions, the equation can be simplified:

$$E_{K+} = -\,61\,\text{mV} \times \log\left(\left[K^+\right]_i / \left[K^+\right]_o\right)$$

We will skip the derivation here and establish some useful basic rules instead:

- The only variables in the equation are the intra- and extracellular concentrations of the ion of interest. Therefore, the equilibrium potential results solely from the unequal distribution of the ions.
- For negative ions (chloride Cl^-; bicarbonate HCO_3^-) the sign of the Nernst equation is reversed.
- For multivalent ions, the valence must be taken into account. In the case of calcium (Ca^{2+}), the value must therefore be divided by 2.
- Passive flows of ions (e.g., currents through ion channels) are always directed in such a way that the membrane potential approaches the equilibrium potential for the respective ion.
- The strength of the electrochemical gradient and the direction of flow of an ion is always determined by the difference between the currently existing membrane potential and the equilibrium potential, i.e., $U_m - E_{ion}$ (see example below).
- When channels are permeable to multiple ions (e.g., non-specific cation channels) or when multiple channels with different ion selectivity are open, a mixed equilibrium potential results, which is determined by the relative membrane permeability for the respective ions. This applies, for example, to the

calculation of the resting membrane potential. The corresponding extension of the Nernst equation is called the Goldman-Hodgkin-Katz equation.

- Active transporters can move ions against the equilibrium distribution under direct or indirect energy consumption. In this case the direction of transport cannot simply be derived from the passive distribution according to Nernst!

Example

Let's take a cell with a "normal" K^+ distribution (outside 4, inside 150 mM) and a membrane potential of -70 mV. Inserting the concentrations into the Nernst equation gives an equilibrium potential of E_{K+} = -61 mV × log(150/4) ≈ -95 mV. The present membrane potential is negative (-70 mV), but still 25 mV more positive than the equilibrium potential for K^+ (-95 mV). Only this difference matters: If potassium channels open, K^+ ions will flow out of the cell because the diffusion along the concentration gradient is stronger than the opposing electrostatic attraction.

If we would hold the membrane potential at -70 mV in a voltage clamp experiment (Sect. 3.1.1), we would observe an outward current of positive ions when potassium channels open. Outward currents are plotted upwards in graphs. If instead we allow changes of the membrane potential in a current clamp experiment (Sect. 3.1.2), the outward flow of K^+ ions would make the cell more negative. It would therefore hyperpolarize, which is represented graphically as a fluctuation of the membrane potential downwards. The most extreme value that the membrane potential can reach in this situation is the equilibrium potential E_{K+} (≈ -95 mV). Here, the equilibrium between the electrically driven current and the diffusion current is reached, so that there is no net flow of K^+ anymore. In reality, a cell is rarely conductive for only one type of ion, so the equilibrium potential would probably not be reached in our example. ◄

2.1.2 Electrical Properties of Cells

2.1.2.1 Passive Properties of Cells: Resistance and Capacitance

We consider a cell in rough approximation as a sphere separated from the outside world by a membrane. Forgive us, dear cell biologists and biochemists, but this simple model often proves useful for us electrophysiologists. The hydrophobic membrane obstructs the flow of hydrophilic particles, including ions. Therefore, it has a high electrical resistance R, which we more precisely call the **membrane resistance** R_m. For ionic currents (I) to flow through the membrane, a driving voltage U is required:

$$U = Rm \times I(\text{Ohm's law})$$

Different cells have very different membrane resistances, ranging from about 10 to several hundred MΩ (megaohm, i.e. 10^6 Ω, where ohm [Ω] = volt/ampere [V/A]).

When ion channels open, the resistance R decreases. Intuitively, this is better seen as an increase in conductance G—which is simply the reciprocal of R:

$$G_m = 1/R_m$$

This makes Ohm's law:

$$U \times G_m = I$$

This simple equation is the **basis of the voltage clamp measurements**, which we will discuss in Chap. 3. If you keep the voltage U (i.e., the membrane potential) constant, then the current I is proportional to the conductance G. The current thus represents the opening and closing of ion channels, so that their behavior can be studied in detail biophysically.

With changes in voltage, a second important electrical property of membranes comes into play, the membrane capacitance. Imagine that the membrane potential of a cell is to be rapidly changed from -70 mV to 0 mV in a voltage clamp experiment. To do this, positive ions must flow to the inside of the membrane or negative ions must be removed from there. The larger the cell—and thus the membrane surface area—the more ions must be moved. A current is therefore needed to charge the membrane surface—only then is the new voltage of 0 mV reached. This charge-storing property is called electrical capacitance which depends on the size, shape, and material of an object. In general, it can be defined as:

$$C = \Delta Q/\Delta U$$

Q is the amount of charge, U is the voltage (potential difference, such as the membrane potential). The capacitance thus describes how much charge must be moved onto an object to cause a given change in voltage. It has the unit coulomb/volt (C/V), which is called farad (F). Voltages do not change instantaneously, so they can be more accurately described as the rate of change over time dU/dt. Then ΔQ becomes the temporal change of the charge dQ/dt, which is nothing else than the current I. To change voltages (membrane potentials) a charging current or "capacitive" current I must flow, which shifts the required amount of charge to the membrane. The formula is then:

$$C = dQ/dt/dU/dt = I/dU/dt$$

By rearranging, we get the charging current:

$$I = C \times dU(t)/dt$$

Thus, the faster the voltage change, the larger is the charging current. For electrophysiological measurements, the capacitance of the cell membrane (C_m) is relevant, but also the capacitance of the pipette or objects in the vicinity of the pipette, such as silver wire, pipette holder etc. Often these are simply summarized as C_{pip}. Membrane capacitance depends on the size and geometry of the cell and typically ranges from about 10 to several hundred pF (picofarad; p = pico = 10^{-12}). Pipette capacitance should be kept as low as possible (Sect. 5.1.4) and should be well below 10 pF.

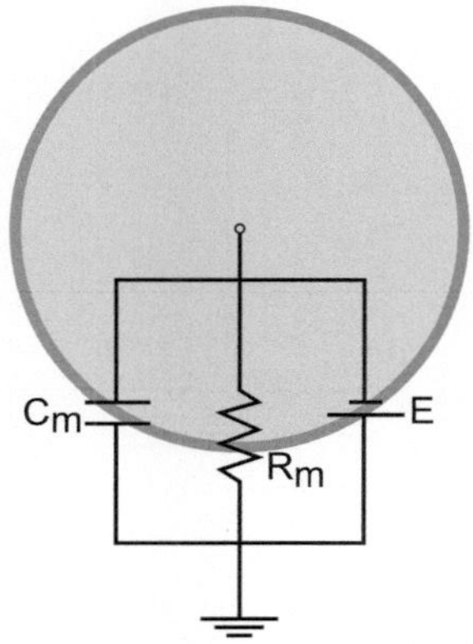

Fig. 2.1 Equivalent circuit diagram of a cell with membrane capacitance (C_m), membrane resistance (R_m) and a battery voltage (E)

We have now described three basic electrical properties of the cell membrane: It has a resistance R_m, an electrical capacitance C_m, and it contains the sodium–potassium-ATPase, which leads to the unequal distribution of ions and causes the negative membrane potential. This primary active transporter/pump can be seen as a kind of battery or voltage source. These properties bring us to the equivalent circuit diagram of the cell shown in Fig. 2.1.

2.1.2.2 Active Properties: Action Potentials

Many of the cells which we study in electrophysiology are excitable, i.e., they generate action potentials. The shape and sequence of action potentials are very variable and are often characteristic for the respective cell type. They are called "active properties" and together with the "passive properties" (R and C) they form the so-called "intrinsic electrical properties" of the cell.

In order to observe action potentials, one must allow for voltage changes, i.e., one does not measure in the voltage clamp mode, where the voltage is specified by the experimenter. This alternative recording mode is called 'current clamp' (Sect. 3.1.2). In this mode, voltage changes of the membrane are recorded, while defined currents can be injected into the cell through the pipette.

The exact shape of the action potential of a cell depends on the expression of ion channels, especially those that are activated by voltage changes. This also applies to the sequence of several action potentials during sustained depolarization. The waveform of an action potential or the frequency of a series of action potentials is thus a kind of "signature" of the cell, which can be used to distinguish different cell types, such as pyramidal neurons, fast-spiking interneurons, cardiac muscle cells or olfactory sensory cells. This is often done by injecting de- and hyperpolarizing currents according to a fixed protocol at the beginning of a recording in current clamp mode. The response of the cell to these currents reveals the passive and active membrane properties and tells us a lot about the cell type—but also about the quality of the recording. Instead of a lot of theory, we show two typical examples in Fig. 2.2.

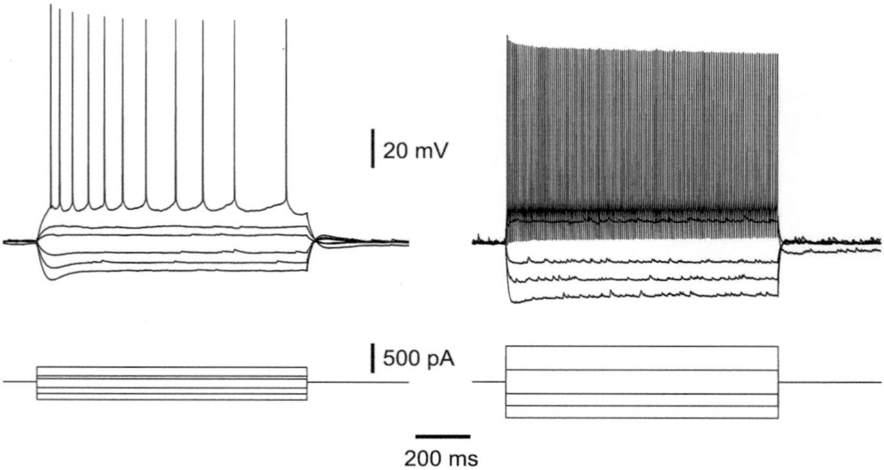

Fig. 2.2 Passive and active properties of a pyramidal cell (left) and a fast-spiking interneuron (right). The voltage responses (top row) were generated by current pulses (pulse duration 1 s) of first negative and then positive amplitude (bottom row)

2.2 Recording Methods (Overview)

Electrophysiological recordings take place on different scales and are used to determine various physical and biological processes. The patch clamp method is just one of many possibilities. It can be used at the level of single cells or subcellular structures down to individual ion channels, and its outstanding feature is the particularly accurate biophysical characterization of membrane currents. In this section, we present most common electrophysiological methods. This is intended to provide an overview, to put the patch clamp technique into a wider context, and to know about alternative or complementary methods if necessary. The main characteristics of the different methods are summarized in Table 2.1.

Beforehand, we should point out that most electrophysiological recordings measure one of two physical parameters: voltage or current. Often the other parameter is controlled by the amplifier. If currents are measured at a given voltage, this configuration is called voltage clamp, if voltages are measured at given currents, it is called current clamp. Measurements of other electrophysiological quantities, such as membrane capacitance (Sect. 6.6) or magnetic fields (see below), are less common.

Therefore, before making a recording, we should know whether we are interested in voltages or currents, and in which order of magnitude between individual ion channels and whole organs we want to measure.

Table 2.1 Important electrophysiological measurement methods and their application

Technique	Level	Measured variable	Properties	Typical Application
Patch clamp	Cell, subcellular	Currents (Voltage clamp) Membrane potential (Current clamp)	Very high resolution, biophysically precisely controlled conditions, subsequent staining of the cell possible	Characterization of ion channels Electrical behavior of individual cells
Intracellular recordings (conventional)	Cell	Membrane potential (current clamp) Rarely also currents (voltage clamp)	Measurement of the electrical behavior of cells under more natural conditions, subsequent staining of the cell possible	Electrical behavior of (subsequently identified) neurons, sensory or muscle cells
Juxtacellular recordings	Cell	Voltage (extracellular)	Recording of action potentials of individual neurons, subsequent staining of the neuron possible	Behavior of individual, subsequently identified neurons in the network
Unit recordings	Cell, network	Voltage (extracellular)	Recording of action potentials of multiple neurons, No staining of cells possible, but classification of individual cells by their electrical signature	Behavior of many individual neurons, (heart) muscle cells, etc. in the network
Field potentials	Network, tissue, organ	Voltage (extracellular)	Summated potentials of local cell networks Synchronization, spatial and temporal patterns	Activity of native neuronal networks, functional characterization of synaptic connections
EEG, MEG	networks, brain	Voltage (extracellular)	Noninvasive recording of coordinated neuronal activity, good temporal resolution, poor spatial resolution	Neuronal correlates of vigilance and behavior/cognition Diagnostics (epilepsy, sleep disorders, disorders of consciousness)
ECG, EMG, ENG	Organ	Voltage (extracellular)	Noninvasive recording of summated action potentials (heart, muscle, nerves)	Clinical diagnostics of electrical excitation processes in the organs

2.2.1 Cellular and Subcellular Recordings

Several techniques are used to measure electrical phenomena in single cells: patch clamp, conventional intracellular recordings, juxtacellular recordings, and unit measurements. The first two methods directly measure electrical processes at the membrane with high resolution, while the latter two are based on extracellular electrodes and capture only the large voltage fluctuations of action potentials.

The **patch clamp technique** is characterized by a very high seal resistance between the tip of a glass pipette and the cell membrane, the gigaseal. This minimizes leak currents and thus noise. The patch clamp technique allows the registration of minute currents down to the tiny ionic currents through single ion channels (Fig. 2.3). The electrical processes of the entire cell can also be measured at high resolution in whole-cell mode. Both pipette and bath solution are defined by the experimenter, such that the ion ratios on both sides of the membrane are predetermined, creating a biophysically well-defined, but rather artificial situation. The technique is suitable for current measurements in voltage clamp mode as well as for voltage measurements in current clamp mode. The latter are practically only performed in the whole-cell configuration.

The patch clamp method is used in vitro on cultured cells or in isolated tissue preparations (e.g., brain slices), and even in vivo by specialized laboratories (Sect. 6.8). Due to the direct access to the cell, substances can easily be introduced into the cell, such as dyes for later structural analysis or for functional imaging, but also drugs for experimental interventions (Sect. 6.11). Figure 2.4 shows two

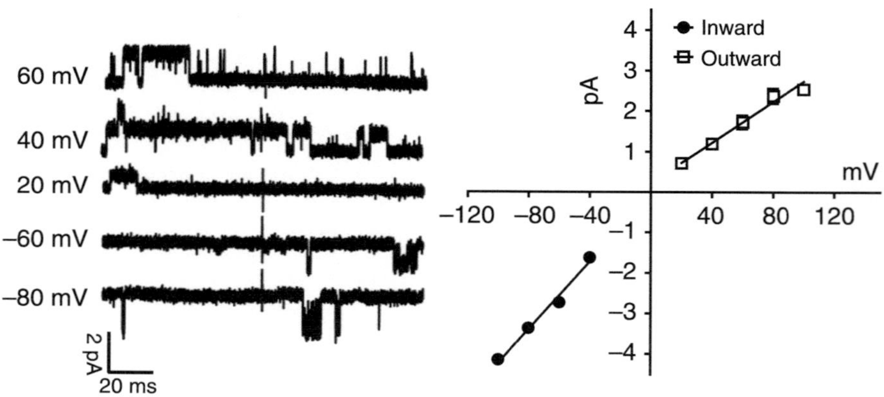

Fig. 2.3 Single-channel currents. Original registrations of the current through individual ion channels in an outside-out recording (left). For this purpose, mechanosensitive PIEZO1 channels were expressed in a cell line and activated by applying pressure to the pipette holder. Recordings at different holding potentials in the voltage clamp configuration produced the current–voltage curve averaged from several patches (right). In this case, it shows a tendency towards "rectification", i.e. outward currents at positive pipette potential are smaller than inward currents at negative potential. The conductance of the channel corresponds to the slope of the curve (dI/dU) at each potential. (From Moroni et al. 2018)

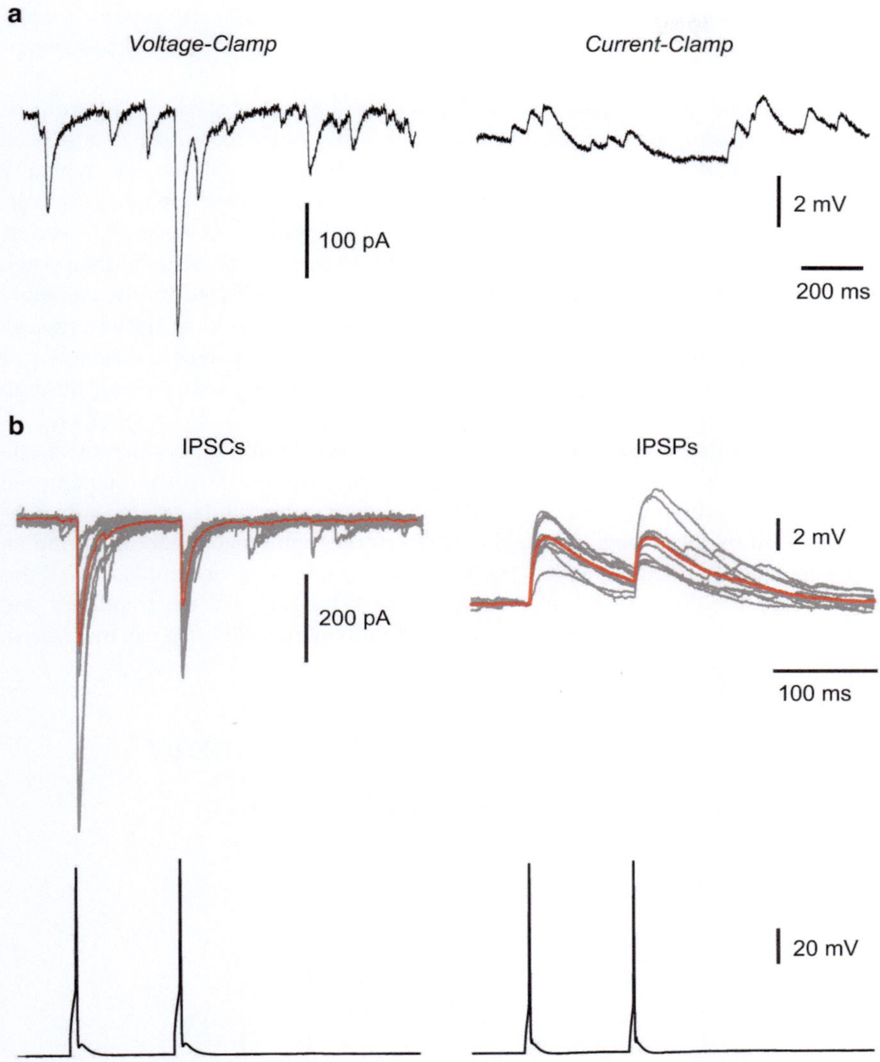

Fig. 2.4 **a** Spontaneous synaptic currents in voltage clamp mode (left) and synaptic potentials in current clamp mode (right) during a whole-cell recording. **b** Inhibitory postsynaptic currents (IPSCs, top row left) and inhibitory postsynaptic potentials (IPSPs, top row right) after triggering action potentials in an upstream (presynaptic) interneuron in current clamp mode (bottom row). In this case, the pipette solution contained a strongly increased chloride concentration, which is why IPSCs are inward currents and IPSPs have a depolarizing effect (Sect. 4.4.5). The averaged responses in the top row are shown in red.

typical measurements of synaptic transmission, one in voltage clamp mode (measuring postsynaptic currents) and one in current clamp mode (measuring postsynaptic potentials).

Conventional **intracellular recordings** (Fig. 2.5) are based on inserting a fine, electrolyte-filled glass pipette into a cell. This older technique is still used to record changes in the membrane potential with high resolution, i.e. typically in current clamp mode. With suitable amplifiers, current measurements in voltage clamp mode are also possible, but these do not provide as noise-free data as the patch clamp technique. However, the configuration is less artificial than patch clamp recordings because the intracellular milieu is less affected by the electrode solution. The intracellular recording technique is mostly used in tissue preparations like brain slices in vitro or in vivo to capture the native electrical behavior of cells. Substances for subsequent staining can be introduced into the cell through the very fine glass pipettes.

For **juxtacellular recordings** (Fig. 2.6), the open tip of a glass microelectrode is placed in close proximity to the cell membrane, without directly touching the cell. In this configuration, it is possible to record extracellular voltage fluctuations evoked by the strong and rapid changes in the membrane potential during an action potential. These so-called ‘spikes’, ‘unit discharges’ or simply ‘units’ are captured by the pipette, so that the “discharge behavior”, i.e. the frequency and temporal pattern of the action potentials of individual cells, can be monitored.

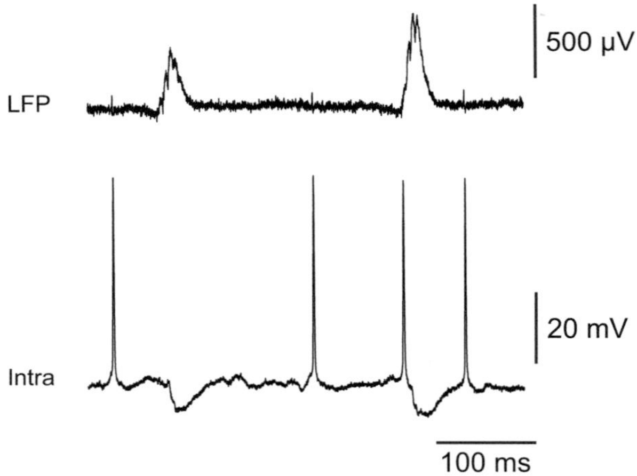

Fig. 2.5 Parallel recording of field potentials and intracellular potential. Extracellular (upper trace) and simultaneous intracellular recording (lower trace) from a mouse brain slice. Extracellular field potentials represent coordinated activity in the neuronal network. The intracellular recording of the membrane potential of a pyramidal cell shows hyperpolarization when the network is activated (here in the form of two so-called *sharp wave-ripple* complexes). Recordings of this type make it possible to characterize the behavior of individual neurons within a network and provide information on the underlying mechanisms (here: synaptic inhibition). (Image courtesy of Alexei Egorov, Heidelberg)

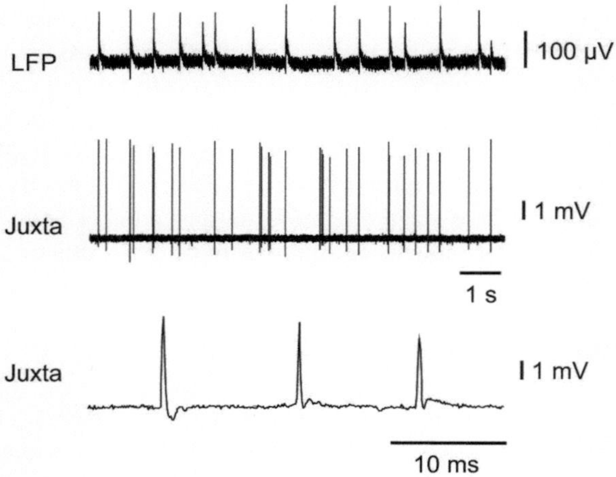

Fig. 2.6 Simultaneous field potential and juxtacellular recording. Here again, *sharp wave-ripple complexes* were measured in a mouse brain slice (upper trace; the data are shown in a more time-compressed form than in Fig. 2.5). The lower traces (with different time resolution) show action potentials of an unidentified neuron in a juxtacellular recording. (Image courtesy of Alexei Egorov, Heidelberg)

Subthreshold signals, i.e. voltage fluctuations below the level of action potentials, are usually not (or almost not) detectable. Given a stable waveform of recorded "unit discharges", one can be relatively sure to record from one individual cell directly in front of the pipette. By injecting current into the pipette, dyes can be "injected" from the pipette solution into the cell, probably through electroporation. This allows the cell to be stained and morphologically classified after the experiment, similar to intracellular measurements. The juxtacellular technique is mainly used in vivo (rarely also in vitro) to measure the spiking behavior of individual, defined neurons in neuronal networks.

Extracellular recordings from single cells can also be performed more easily, but then one has to give up the possibility of staining and thus the morphological identification of the cells. Instead of using the fine, hollow glass electrodes of the juxtacellular method, simpler electrodes made of thin tungsten wires (as single or multiple electrodes) are used. They allow relatively easy positioning and long-term recordings of action potentials (units) in the brain of freely moving animals, or even in humans in certain clinical conditions. There are also multi-electrode arrays for in-vitro experiments, which can measure action potentials from a large number of spatially distributed cells (e.g., in a brain slice preparation or in a culture of heart muscle cells grown directly on the array). All these techniques have in common that the individual cells from which the action potentials originate are not visible and cannot be subsequently identified by staining. However, with multiple, closely positioned electrodes it is possible to distinguish different electrophysiological signatures of different units (cells), depending on the distance of the respective cell to the individual electrodes. The distinction between multiple units requires considerable effort and experience. Such data sets are referred to as identified units in contrast to simple multiunit activity with multiple action potentials from unidentified, presumably different sources. A commonly used electrode type are "tetrodes" made of four closely spaced, twisted wires, which allow recordings of different signatures from several differently positioned cells. By now, there are also commercially available multi-electrodes in any number and shape, allowing for the parallel recording from tens or hundreds of different units.

2.2.2 Summated Extracellular Potentials

At the next level, we are no longer concerned with the electrical contributions of individual cells, but of cell populations. In this case, the extracellular contributions of individual action potentials or synaptic potentials add up and generate small (μV to mV) extracellular potential fluctuations, which reflect the averaged behavior of the cells.

Locally restricted extracellular potentials are referred to as **local field potentials (LFPs)**. They are measured, for example, with metal or glass electrodes in brain slices or in vivo. The amplitude of potential fluctuations not only reflects the size of the contributions of individual cells, but above all their synchrony. For example, if many axons in a neural network are activated simultaneously, they will trigger synchronous postsynaptic potentials in their target cells. The activities of the individual cells add up to a uniform, macroscopically visible wave, that is clearly distinguishable from baseline. If, on the other hand, many neurons are completely uncoordinated in their activity, the postsynaptic potentials only form an unspecific noise. Summated field potentials are therefore a measure of the synchrony in a network or tissue. Field potential recordings are mainly used in neurophysiology to measure the strength, but also the direction, spatial spread and the temporal behavior of synaptic potentials. With experimental stimulation of larger axon bundles or during hypersynchronous activity (epilepsy), it is also possible to record summated action potentials (population spikes). Action potentials of individual cells (units) can be isolated from good field potential recordings as described above. For this purpose, the recorded data are filtered with a so-called high-pass filter, which eliminates all slow signal components (e.g., below 500 or 1000 Hz). The remaining fast, short impulses are units, which may be correlated with the original field potential to understand the behavior of individual neurons in the network. In addition to simple, homemade wire or glass electrodes, the multi-electrode arrays, tetrodes etc. already described for units can also be used to measure field potentials.

Signals measured noninvasively in animals or humans using **electroencephalography** (EEG) or **magnetoencephalography** (MEG) have even greater spatial extent (Fig. 2.7). In EEG, electrodes are placed allover on the scalp to record summated potentials emanating from the upper cortical layers. The MEG does not directly measure electrical activity, but rather tiny magnetic fields that result from

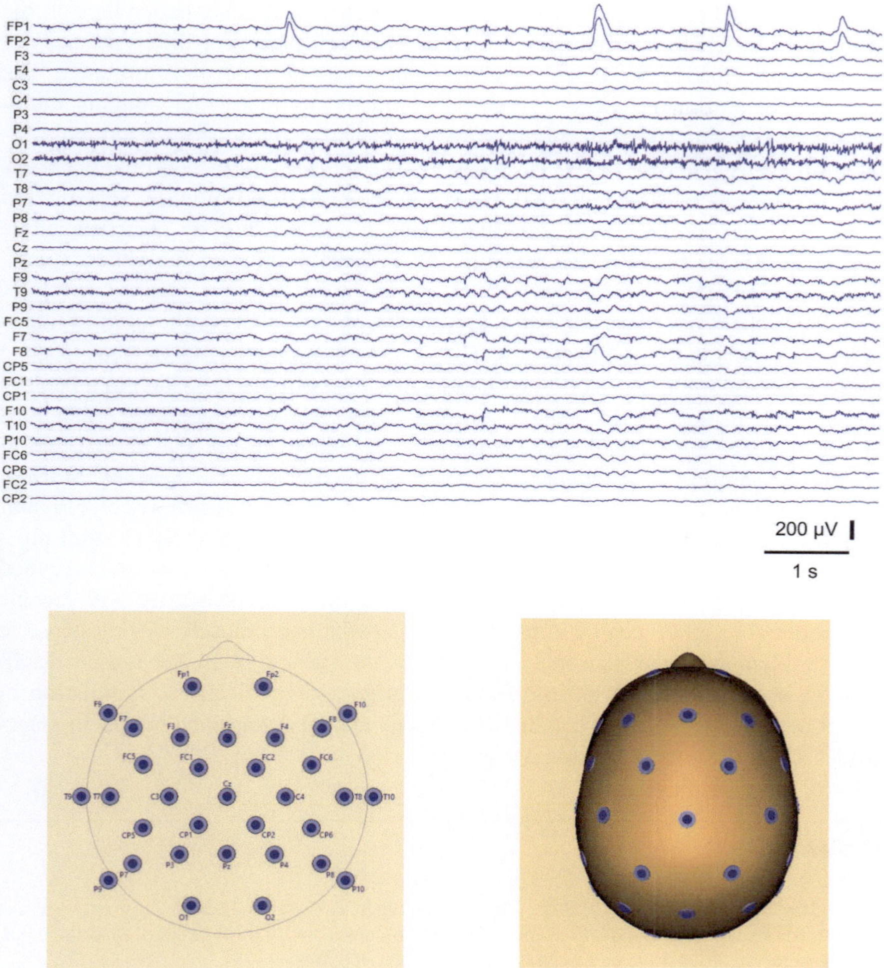

Fig. 2.7 EEG measurements. Section of an electroencephalogram (top). Characteristics are the high number of parallel recorded channels (here: 33), the occurrence of rhythmic activity (e.g. in F9, T9, and P9 shortly after half of the displayed period), and the overall low amplitude of < 200 µV. In the two upper traces (FP1, FP2), four artifacts can be seen due to eye blinks. The lower schematics show the positions of the recording electrodes. (Courtesy of Steffen Syrbe, Heiko Kienzler and Andre Rupp, Heidelberg, Germany)

currents along the dendrites of the parallel oriented cortical neurons. The structure of the data resembles the EEG, but it has better spatial resolution. MEG measurements are technically demanding and are only performed in specialized centers, while EEG measurements are inexpensive, easy to perform and widely used. These two methods essentially measure synchronous synaptic activity (action potentials are never sufficiently synchronous in the healthy brain to be visible in the EEG). The cortical networks show different oscillations depending on the activity state, which can be used to distinguish sleep phases or different states of attention, but also pathological synchronizations as they occur in epilepsy. The temporal resolution is good—especially compared to functional imaging methods—and is about 30 ms in routine measurements (resulting from a low-pass filter limiting frequencies to ~ 30 Hz). This allows the activity of neuronal networks to be recorded almost in real time. However, the spatial resolution is very limited (~ cm), especially since only superficial signals can be directly recorded, while events in deeper brain areas can only be indirectly inferred (so-called source localization).

Finally, there are three clinical diagnostic techniques that can be used to measure action potentials noninvasively. The electrocardiogram (ECG) measures the propagation of summated action potentials in the heart muscle, which in a healthy person runs in an orderly fashion, forming large, stable extracellular potentials. Gradients between excited and unexcited areas form during the cardiac activity cycle, resulting in a sequence of potential fluctuations that provide information about the course, duration and spatial organization of atrial and ventricular activity. The electromyogram (EMG) measures action potentials in skeletal muscle, where spikes from individual muscle cells or small groups of myocytes (called motor units) can be detected, especially when recorded invasively with needle electrodes. The EMG can be used to determine whether a muscle weakness is due to a problem with the neuronal "control" of the muscle or in the muscle itself. Finally, summated action potentials can be triggered by electrical stimulation of larger nerves and the nerve conduction velocity can be determined by placing electrodes along the nerve (electroneurography, ENG).

References

Covey E, Carter M (Eds.) (2015) Basic Electrophysiological Methods. Oxford University Press.

Moroni M, Servin-Vences MR, Fleischer R, Sanchez-Carranza O, Lewin GR (2018) Voltage gating of mechanosensitive piezo channels. Nat Commun 9:1096

Technical Basics of the Patch Clamp Method

3

3.1 Operating Principle of a Patch Clamp Amplifier

In this chapter we want to discuss the biophysical and technical basics of the patch clamp technique, as far as they are relevant for the experimental procedures and the understanding of the obtained data. For this, no in-depth knowledge of physics or electrical engineering is required. Rather, a few simple principles are sufficient to understand the functions of the amplifier to the extent that one can perform recordings correctly, understand data and identify problems. This, in turn, facilitates the often necessary "troubleshooting"… The practical implementation of the actual experiments will be discussed in Chap. 5.

Using the patch clamp technique, it is possible to measure both currents (at given voltage, voltage clamp, VC) as well as voltages or membrane potentials (at a given current, current clamp, CC). We will first describe the voltage clamp mode, in which the membrane potential is "clamped" to predefined values and the currents required to achieve this are measured (see also Sect. 2.1.2). In the simplest case, the relationship between voltage and measured current is described by Ohm's law as a simple proportionality. The relevant voltage U (figuratively one could say "driving force") is not simply the membrane potential, but the distance of the actual potential from the reversal potential for those ions whose flows we are measuring. An illustrative example of this can be found in Sect. 2.1.1. With this explanation for U, the following applies:

$$U = R_m \times I(\text{potential} = \text{membrane resistance} \times \text{current flow})$$

Instead of the resistance R_m we use the reciprocal $G_m = 1/R_m$, which is called the conductance and intuitively reminds us of the electrically conducting ion channels in the membrane. This results in:

$$U \times G_m = I(\text{membrane potential} \times \text{conductance} = \text{current flow})$$

F. C. Roth et al., *Mastering the Patch Clamp Technique: A Practical Guide*,
https://doi.org/10.1007/978-3-662-70745-6_3

Thus, I is proportional to the conductance G at a given voltage. This is exactly what voltage clamp measurements are all about: We want to measure changes in the conductance of the membrane and we use the injected current as a measure. Often, one tries to create a situation where only a very specific type of ion channel opens or closes. The compensating current then provides a precise biophysical description of the function of each channel. In this way, for example, the effect of a neurotransmitter on postsynaptic receptors or the amplitude and kinetics of voltage-activated ion channels in heart cells can be studied.

3.1.1 Voltage Clamp

The first voltage clamp amplifiers used two separate pipettes for voltage measurement and current injection. One reason for this is that the pipettes themselves have a resistance. If the current is injected through the same pipette which is also used to measure the voltage, a voltage change will occur at the pipette resistance (R_{pip}) according to Ohm's law. Thus, the amplifier no longer "sees" the membrane potential, but the sum of the potential changes at the pipette and the membrane. The higher the injected current, the more the measurement would be distorted. This problem can be solved by injecting the current through a separate pipette. However, the two-electrode voltage clamp (TEVC) technique can naturally only be applied to large cells, such as neurons of some invertebrates or *Xenopus* oocytes. Therefore, techniques have been developed in which voltage measurement and current injection can be performed with the same pipette and still provide reliable data. The patch clamp technique is the most widely used and most sensitive of these single-electrode voltage clamp (SEVC) methods. How does it work? The main "trick" is shown in the simplified circuit diagram in Fig. 3.1a.

The circuit shown in Fig. 3.1a represents a current-to-voltage converter. It is located in the pre-amplifier, as close as possible to the pipette, to avoid noise and signal losses due to long cables (Sect. 4.3.1). Central elements are the "operational amplifier" (OPA) and the feedback resistor R_f to its inverting input (in some models, a capacitor C_f is used instead of R_f). The OPA is a complex electronic component that looks like a small black box with 8 pins, and fortunately can also be considered a "black box" by the user. We propose four simplified rules:

1 At the output (tip of the triangle symbol) an OPA delivers an amplified voltage that depends on the difference of the input voltages. By this, it acts as a voltage-controlled voltage source.
2 The two inputs of an OPA detect voltage without any significant current flow. Therefore, the OPA has a very high (almost infinite) "input resistance".
3 In a feedback circuit as shown in Fig. 3.1a, the output of the OPA constantly provides the necessary current to minimize any voltage differences at its inputs.
4 The voltage difference across the feedback resistor R_f is proportional to the current flow between point 1 and point 2 in Fig. 3.1a. This rule is a simple application of Ohm's law.

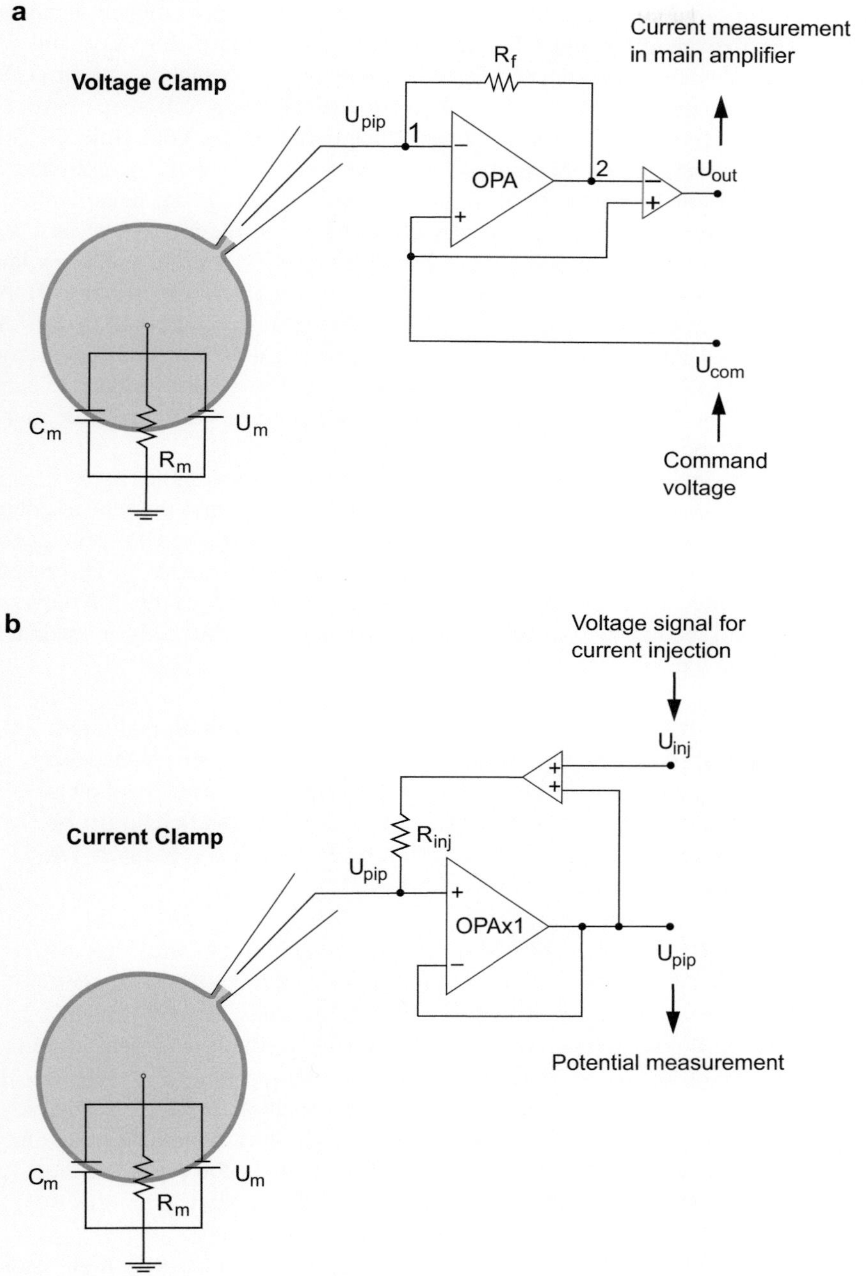

Fig. 3.1 Simplified schematic circuit diagram of a patch clamp pre-amplifier with cell. **a** Voltage clamp circuit **b** Current clamp circuit. OPA: operational amplifier

From this information, the principle of the voltage clamp measurement becomes clear: The two inputs of the OPA are the potential of the pipette U_{pip} and the experimenter's target potential (command potential) U_{com}, which is pre-set at the amplifier or in the control program (Sigworth, 1995b). Any difference between U_{com} and U_{pip} forms an amplified voltage at the output of the OPA (rule 1). This immediately establishes a potential difference between point 1 (U_{pip}) and point 2, creating a current flow through R_f until U_{pip} and U_{com} are equal again (rule 3). The current through R_f flows practically completely into the pipette, because the OPA has an "infinitely" high input resistance (rule 2). According to Ohm's law, a voltage now forms, which is proportional to the current: $U_f = R_f \times I$ (rule 4). As a results, U_{pip} is kept equal to U_{com} and the voltage at R_f is proportional to the current injected into the pipette (hence this circuit is called a "current-to-voltage converter"). With this voltage, we have a measure for the current needed for compensation of any conductance changes in the cell, i.e. we basically measure membrane conductance.

The above processes occur almost in real time (~ femtoseconds $= 10^{-15}$ s), so that initial deviations are corrected practically immediately and the pipette potential is always equal to the command potential. This is because the OPA provides an extremely large amplification of the difference between U_{com} and U_{pip}. This results in a steeply increasing correction current at the slightest deviation between the command and the actual voltage, so that the deviation is practically immediately eliminated via the feedback loop (R_f).

▶ **Caution:** This optimistic statement is only approximate. In real measurements, some complications occur, especially when measuring large, complex cells or when the pipette resistance is not negligible compared to the membrane resistance. These complications must be known and taken into account so that no nonsense is measured. We will discuss this in Sect. 3.2 and in Chap. 4.

The output voltage of the OPA is sent to the amplifier's control unit where it serves to infer the current actually injected into the pipette, using an appropriate calibration factor (depending on the size of R_f). However, at point 2 of Fig. 3.1a, there is not only the voltage across R_f, but also U_{pip}, which is present at point 1 and is added to the current-induced voltage. Therefore, before reading out the signal taken for the current, U_{com} (as an approximation to U_{pip}) is subtracted (Fig. 3.1a, differential amplifier directly before U_{out}). Thus, both tasks of a voltage clamp amplifier are fulfilled: The potential of the cell (or the pipette) is equal to U_{com} and the necessary compensation current is measured via U_{out}. In this way, all changes in membrane conductance can be directly observed.

The particular strength of the patch clamp method lies in its very high resolution, which allows even for the observation of currents through single ion channels. Electrophysiology has thus arrived at the molecular level. Currents can be measured in the picoampere (pA, 10^{-12} A) range, which is about ten billionth of the current flowing through a typical light-emitting diode. For example, the brief opening of a single ion channel for 1 ms at a current amplitude of 10 pA would be caused by less than 10,000 individual ions. The pre-amplifier is optimized for these tiny currents—it operates at very low noise levels and contains high-quality feedback resistors of up to 50 GΩ (gigaohm, 10^9 Ω), which generate well measurable voltages even at the smallest currents. The high input resistance of the OPA is also an important prerequisite for correct and low-noise measurements. At least one manufacturer (Axon Instruments/Molecular Devices) uses a capacitor (C_f) instead of the feedback resistor (R_f), which is even actively cooled in some models. This leads to even lower noise and is therefore an advantage in single-channel measurements. However, as current flows into the capacitor, an increasing voltage builds up that must be periodically "zeroed" to stay within the control range. For larger currents in the whole-cell configuration, these amplifiers are therefore switched back to the conventional feedback with resistors, since they would otherwise have to discharge C_f too often. In this situation, noise is dominated by other sources anyway.

3.1.2 Current Clamp

A fundamentally different configuration is the current clamp mode. Here, the membrane potential is allowed to change freely, as it corresponds to the natural electrical behavior of cells. The circuit required for this is not intended to stabilize the voltage, it only measures it. Almost all patch clamp amplifiers have at least one circuit for current clamp measurements, but only certain models are suitable for highly precise measurements. This should be taken into account when making your selection, so, pay special attention if you are planning such measurements.

For the actual measurement, you need an OPA without amplification function (× 1) but with high input resistance. This ensures that the measured membrane potential is not distorted by significant current flow into the electronics (Fig. 3.1b). As a second component, you also need a current injection device, with which you can depolarize or hyperpolarize the cell at will. Instead of a command voltage, you have a command current that can be set via the main amplifier or the recording software. This current is generated by adding a voltage to the measured potential of the cell and passing it through a resistor into the pipette (Fig. 3.1b). According to $I=U/R$ (Ohm's law), the current amplitude depends on the additional voltage and the built-in resistor. Since this value is known, the command signal at the amplifier can be directly calibrated as a current command. However, the current also causes a voltage at the pipette, because it has a resistance that reacts to the current flow according to $U=R_{pip} \times I$. This can distort the measurement, the more so the more current is injected. This additional voltage is compensated by generating an equal signal via a variable resistor, which is then subtracted from the measured potential. In reference to a similar, historical circuit, this is called the Wheatstone Bridge or simply "bridge balance". The corresponding signals are shown in Fig. 3.3c.

Current clamp recordings are of great importance whenever the native electrical behavior of cells is to be assessed. This can be the case for electrically excitable cells that have been differentiated from pluripotent stem cells, or for the analysis of the natural behavior of neurons within their network. With the patch clamp method, current clamp recordings are mainly performed in the whole-cell configuration (Sect. 5.2.4).

In current clamp recordings, there is often a slow drift of the membrane potential, as the membrane resistance or leak conductance changes. One is usually not interested in these changes, but rather in faster voltage fluctuations (postsynaptic potentials, action potentials etc.) at a more or less constant membrane potential. To keep the average membrane potential constant, some amplifiers offer a kind of hybrid of voltage clamp (for slow changes) and current clamp (for fast changes). They go by various names such as slow current injection, low-frequency voltage clamp or dynamic holding (Dolzer, 2021). A typical application of this configuration is the determination of the reversal potential of spontaneously occurring postsynaptic potentials (PSP). For this purpose, a nerve cell can be held for some time at approximately -80 mV, -75 mV, -70 mV etc. and the synaptic events at these potentials can be collected. The fast potential fluctuations of the individual PSPs are practically not influenced by the slow voltage stabilization. The amplitudes of the PSPs can then be plotted against the different average potentials to obtain the reversal potential of the observed synaptic inputs. If, for example, the PSPs disappear at about -70 mV, are depolarizing (positive) at more negative values and hyperpolarizing (negative) at more positive values, we would obtain a reversal potential of -70 mV. This typically indicates a selective conductance for chloride (Cl^-) and would suggest that GABAergic or glycinergic inhibitory synapses are causing these PSPs.

3.1.3 Discontinuous Amplifiers

For voltage clamp recordings using whole-cell mode, a completely different measuring principle has been developed that also works with a single electrode: the discontinuous single-electrode voltage clamp (dSEVC) amplifier. In this case, the compensatory current for setting the target voltage is periodically injected during very short time windows, followed by a pause. During the current injection, a potential is built up at the pipette according to its resistance, which would distort the voltage measurement. During the pause, this potential at R_{pip} drops again. The effect of the current injection into the cell is largely preserved despite the pause, because, due to its large capacitance (Sect. 2.1.2), the cell membrane discharges much slower than the pipette. Thus, after the current injection has ended and the additional potential at R_{pip} has disappeared, the membrane potential can be measured in a current-free, undistorted situation. The membrane potential is now compared with U_{com}, and the next short current pulse is injected to counteract any deviations. In a so-called sample-and-hold amplifier the measured potential U_{mem} is stored and made available during each cycle, so that an apparently continuous membrane potential recording is given at the amplifier output. The crucial technical trick is to bring the current-induced voltage at the pipette resistance back to zero very quickly after each injection—otherwise, one can either measure only extremely slow processes or will get false values for the membrane potential and the compensatory current. This requires an optimized compensation for the pipette capacitance (C_{pip}), which must be precisely adjusted before each measurement (Sect. 3.2.1).

Modern dSEVC amplifiers can operate at frequencies of more than 20,000 Hz so that rapid conductance changes can be easily studied in the voltage clamp mode. Since they are virtually independent of the pipette resistance when properly set, they can be used with both patch electrodes (resistances typically < 10 MΩ) and conventional "sharp" microelectrodes with much higher resistances (up to > 100 MΩ). However, they cause more noise than continuously operating patch clamp amplifiers and are therefore not suitable for the highest resolution (single channels). Some models therefore combine conventional patch clamp amplifiers and dSEVC in a single device.

3.2 Complicating Factors, and How to Deal with Them

In the following sections, we will discuss three factors that may cause artifacts or misinterpretation of data. These "troublemakers" are electrical capacitance, series resistance, and offset potentials. By the end of this chapter, you should understand what these quantities are, what problems they cause, how to recognize them, and how to deal with them in principle. Concrete instructions and tips for practical handling of the error sources can be found in Sect. 5.2.4.

3.2.1 Capacitance

We have already learned the concept of capacitance in Sect. 2.1.2. Simply put, it is the charge that must be applied to an object (e.g., a cell membrane or a pipette) to change its electrical potential by a certain amount. This requires a current to flow that is proportional to the voltage change and to the capacitance (i.e., the charge-storing property of the object). Formally, this can be described as follows:

$$I = C \times dU(t)/dt$$

Or resolved according to the rate of the voltage change:

$$dU(t)/dt = I/C$$

These equations reveal the problem caused by the capacitance: it slows down voltage changes. In fact, the value of dU/dt (voltage change per unit of time) is smaller, the larger the capacitance C or the smaller the current I. If you want to have fast changes (large values of dU/dt), the capacitance should be as small as possible and the charging current as large as possible. Both are not arbitrarily controllable in patch clamp measurements, so the influence of the capacitance cannot be neglected. In addition, capacitive charging currents add to the measured total current, and hence distort the measurements. They increase noise or, in the case of large voltage jumps, induce large and disturbing transient charging currents. However, measurements can be optimized by minimizing capacitances and by using compensation circuits for the remaining C. Relevant capacitances arise at three points: At the feedback resistor of the pre-amplifier, at the pipette and the pipette holder, and at the membrane of the measured cell.

The **feedback resistor** R_f as an electronic component naturally has a finite extension, so that for each voltage change a certain amount of charge must flow onto the resistor. So, it has a capacitance. A rough calculation (Sigworth, 1995a) shows that it reacts to a rectangular voltage jump (e.g. of the cell membrane) with a voltage change that corresponds to an exponentially increasing curve with a time constant τ of about 1 ms (i.e. it shows only 63% of an "instantaneous" current after 1 ms). This is much too slow for the recording of rapidly rising signals, such as single channel currents. For this reason, correction circuits are built into the amplifier that compensate for the signal losses due to the charging of the capacitance. (The principle of such a circuit will be described below using the pipette and membrane capacitance as an example.) This reduces the response time of the pre-amplifier to a few microseconds, which is sufficient for most applications.

The Pipette and pipette holder also have capacitances, which are usually summarized as the pipette capacitance C_{pip}. When the command potential changes, the conductive components connected to the cell must also be recharged to assume the new voltage, e.g. the water-wetted parts of the glass pipette. Typical values of C_{pip} are up to 10 pF (10×10^{-12} F), but preferably much lower. They vary depending on the pipette glass (thick-walled glass has smaller capacitances) and the extent of conductive surfaces (it helps to keep the water level in the recording chamber low and to keep the pipette holder dry; see Sect. 5.3). Since C_{pip} varies depending on the recording setup and the experiment, amplifiers feature a variable compensation for the pipette capacitance. Figure 3.2 shows its function. Basically, the capacitive charging current for the pipette and pipette holder is injected through a separate path, which is not visible in the measurement. For this purpose, the command voltage is passed through a so-called RC element, whose resistance is variably adjustable. When properly adjusted, this RC element generates exactly the time course of the voltage with which the pipette and the pipette holder are charged. This voltage is stabilized in the downstream operational amplifier and differentiated via a so-called injection capacitor C_i ($I = C_i \times dU/dt$; the capacitor can thus be used to differentiate a voltage course). The thereby generated current exactly corresponds to the charging current required for the voltage changes at the pipette. However, it is it is not injected via the current–voltage converter, but via a "side branch". Thus, the current signal of the actual recording circuit appears almost completely smooth, with very little capacitive charging of the pipette being visible (Fig. 3.2b). The amplitude and time constant of the charging current are set

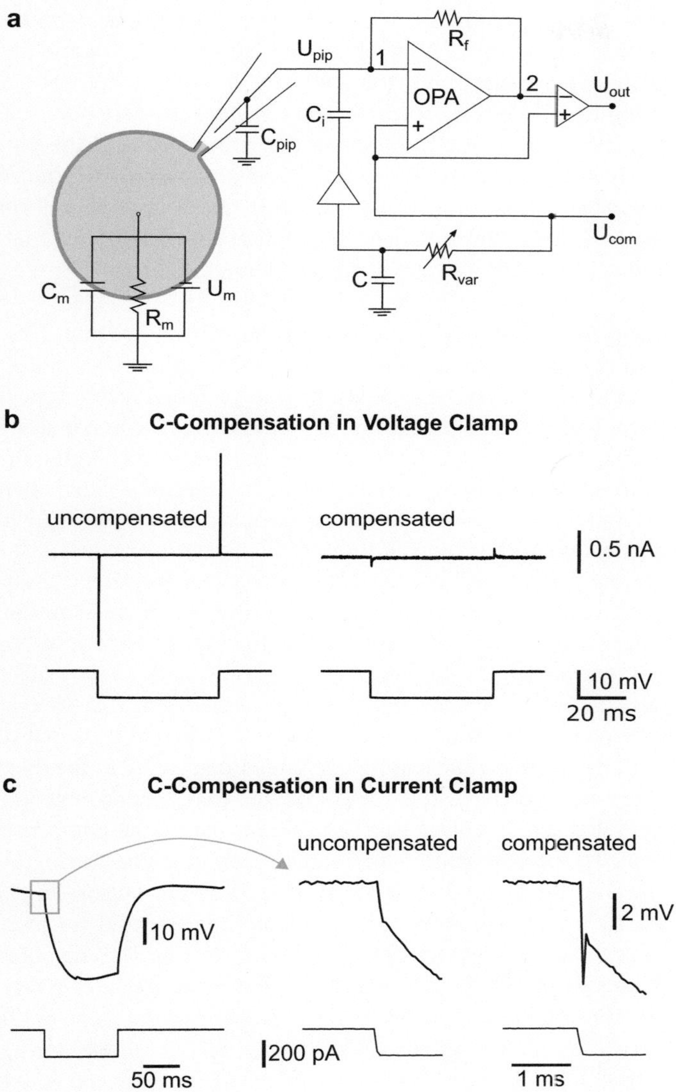

Fig. 3.2 **a** Schematic circuit diagram including C-compensation in the voltage clamp circuit. **b** Changes in the current response to a command voltage pulse after compensation in the voltage clamp. **c** Change in the potential response to a current pulse after compensation in the current clamp mode. Unlike in voltage clamp mode, the compensation results in a small voltage artifact at the beginning and end of the current injection

either manually using potentiometers or via the software of computer-controlled amplifiers. The latter optimize the compensation automatically and output the calculated capacitance as a numerical value.

▶ **Caution:** The compensation of pipette and membrane capacitance means that the corresponding charging currents in voltage clamp recordings do not appear in the "officially" measured current signal. This is the desired effect, but it should not hide the fact that voltage changes still require time to recharge the membrane and the pipette due to their capacitances. Thus, the amplifier's output is reduced to the biologically interesting signal (which is good), but we still might need more time for any voltage changes than desired. It is just that the required capacitive charging current is no longer visible.

The **membrane capacitance** (C_m) must also be taken into account. It is quite significant due to the enormous thinness of the membrane (typical values are up to 100 pF, or even more for very large cells). In voltage clamp recordings, this capacitance delays sudden voltage changes at the membrane as compared to the speed of the command potential (Fig. 3.2b). The time needed for recharging of the membrane may lead to distortions of the measurement, especially when characterizing voltage-gated ion channels. Amplifiers have a compensation for C_m, which works on the same principle as the compensation of the pipette capacitance. It makes the transient charging current largely invisible by bypassing the circuit used for measuring injected currents. However, it does not cause a real acceleration of membrane charging! Exceptions to this rule are amplifiers that work with the so-called supercharging (see below).

Distortions also occur in current clamp measurements due to capacitances. Changes in the membrane potential can only be measured at the input of the OPA as fast as the parts of the pipette and pipette holder connected to the membrane are recharged. This can greatly distort rapid changes (e.g., action potentials) if C_{pip} is not compensated for. In contrast to the voltage clamp, the compensation actually accelerates the measurement, which can be seen in the narrowing of the small voltage artifacts at the beginning and end of test current pulses (Fig. 3.2c). The maximum possible compensation of C_{pip} is therefore essential for recording fast signals in current clamp. The situation is different for C_m: The membrane capacitance contributes to the kinetics of electrical signals in real life, for example by influencing the propagation of a PSP from the dendrite to the soma of a neuron (in the simplest case by slowing down and reducing the amplitude during "electrotonic conduction"). Thus, when the pipette is located at the soma of the cell, one simply accepts the effect of the membrane capacitance in order to correctly measure the real course of the somatic membrane potential. If you want to know what the potential at the site of origin of a PSP looks like, it may be necessary to record directly from the dendrite (which we certainly do not recommend as a beginner's project).

Regardless of all the compensation circuits, you should always minimize the capacitive currents in the experiment. Appropriate measures and tricks can be found in Sect. 5.3.

3.2.2 Series Resistance

In whole-cell mode, the electrical resistance caused by the pipette itself is positioned in series between the operational amplifier (OPA) and the interior of the cell. It is referred to as the series resistance R_s (Fig. 3.3). Its size is mostly defined by the limited diameter of the pipette, resulting in typical resistances of 3–5 MΩ. However, membrane fragments and organelles often obstruct the pipette opening and increase R_s. In practice, the series resistance in whole-cell recordings is therefore significantly higher than the resistance of the open, unused pipette. In publications, the target value of R_s is often stated as not more than twice the pure pipette resistance, but even this is hardly achievable in many situations; one often has to live with three times the pipette resistance or even more. But what are the consequences of the series resistance for the measurements?

Let us assume that the cell has a membrane potential of -60 mV in whole-cell mode. We now want to hyperpolarize it to -80 mV using voltage clamp. To do this, we must continuously inject a negative current that makes the intrinsic potential 20 mV more negative than -60 mV. Our amplifier then shows us -80 mV—the set value and the measured voltage are the same. However, the actual voltage is not measured inside the cell, but at the OPA. The amplifier cannot distinguish between voltage gradients at the resistors of the pipette and the membrane, but simply treats them as the sum of $R_s + R_m$. At such a so-called voltage divider, the voltage drops proportionally to the size of the individual resistances. The -20 mV generated by the amplifier are therefore distributed across R_m and R_s instead of just setting the membrane potential (Fig. 3.3a). This doesn't matter as long as R_s is very small compared to R_m and therefore negligible. But let's assume that R_s is 20 MΩ and R_m is 60 MΩ (depending on the cell type and condition, this is not an unrealistic value). Then a quarter of the voltage, i.e. 5 mV, would drop across R_s and only the remaining three quarters, i.e. 15 mV, would drop at R_m. The actual membrane potential would therefore only be -75 mV, although the amplifier tells us that the full -80 mV have been reached. This information is true, but only for the sum of $R_s + R_m$. On the other hand, if the series resistance is only 5% of the membrane resistance, the error is only 5% or 1 mV. The cell would therefore be held at -79 mV instead of -80 mV, which is usually acceptable. However, even in this case, the ratio of R_s to R_m worsens as the membrane conductance increases (reduction of R_m), so problems can still occur during the course of the recording (see below). A similar error occurs in the current clamp mode, where the injection of a current causes a voltage drop across R_s and thus leads to a distortion of the measured membrane potential (Fig. 3.3c).

The charging of the capacitance is also slowed down by high series resistances—after all, the charging current has to flow through the "bottleneck" of the pipette opening. The lower the series resistance, the more current can flow from the amplifier into the cell to compensate for the difference between the target and actual voltage. R_s also interferes when the conductance of the cell membrane

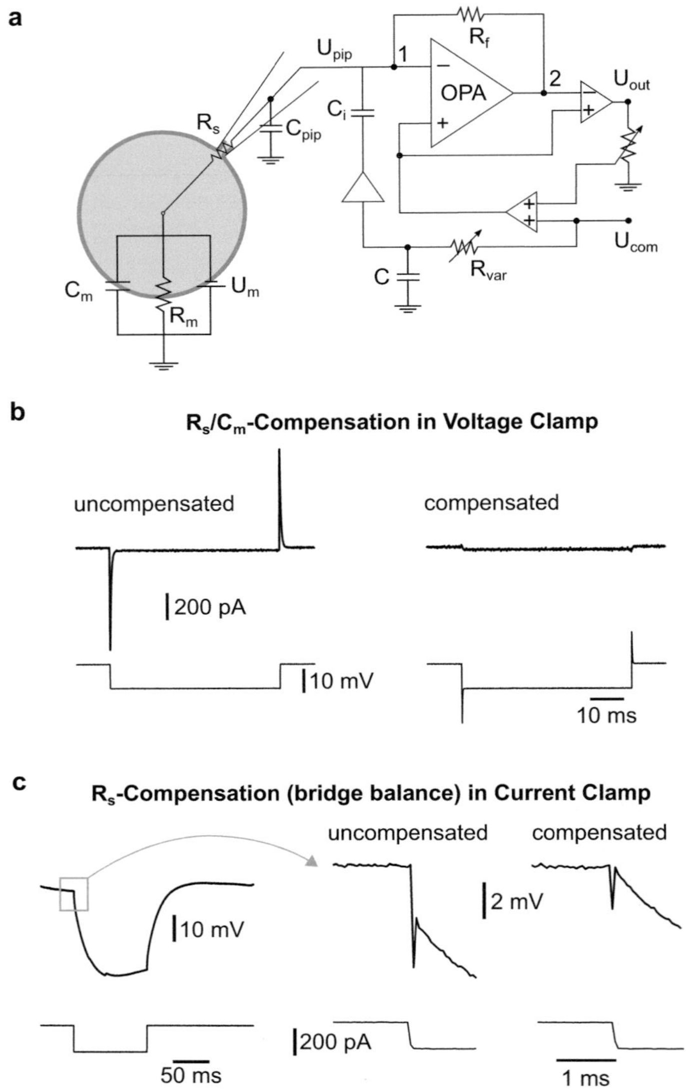

Fig. 3.3 **a** Schematic circuit diagram including R_s compensation in the voltage clamp circuit. **b** Changes in the current response to a command voltage pulse after compensation in the voltage clamp mode. Full compensation is only possible with compact, unbranched cells. **c** Change in the potential response to a current pulse after compensation in the current clamp mode.

changes, for example during a postsynaptic current. In voltage clamp mode, the associated potential change should be compensated immediately by injecting a countercurrent. However, the compensating current is too small when it has to pass through a large R_s (which generates an apparent voltage correction in the desired direction), so that the potential change is compensated incompletely and delayed. The amplitude and kinetics of the measured current will be distorted accordingly. In the worst case, the unwanted change in membrane potential may lead to the activation of voltage-dependent ion channels (especially Na^+ channels), which can cause large and rapid changes in conductance and induce massive voltage deviations (Fig. 5.4).

In summary, series resistance should be kept low. The larger the series resistance relative to the membrane resistance, the more incompletely and slowly the membrane potential is adjusted to the target value, and the more the measured currents are distorted (slowed down and reduced) compared to the actual change in membrane conductance.

3.2.2.1 Compensation of the Series Resistance

The negative effects of series resistance can be partially compensated by adding an additional voltage proportional to injected current to each command voltage. This is done by adding a variable portion of the output signal (corresponding to the injected current) to the command potential U_{com}. In this way, part of the expected voltage loss at the pipette can be compensated for and the desired membrane potential is reached more completely and more quickly. The corresponding circuit is shown in Fig. 3.3a.

The resulting voltage at the pipette is no longer exactly equal to U_{com}, but is increased according to the series resistance (Fig. 3.3b, bottom right). In addition, the transient currents for recharging the cell capacitance become larger when series resistance compensation is enabled, so that voltage changes are implemented more quickly. The combination of C_m and R_s compensation actually improves the speed and accuracy of the voltage clamp method! However, the additional control loop includes a positive feedback: By increasing the output signal of the OPA, the correction of the command potential also increases, which in turn increases the output signal of the OPA. Such positive feedbacks are inherently unstable, and so it is here. With increasing R_s compensation, the noise of the current signal increases and may, beyond a certain threshold, induce uncontrolled oscillations of the amplifier, which usually leads to loss of the cell. Therefore, even under optimal conditions, it is practically never possible to compensate more than 90% of the series resistance, and realistic values are 60–70%.

Also in current clamp mode, a voltage drop at R_s occurs with each current injection. This effect can be seen with rectangular command current pulses, which cause a rapid voltage change at the pipette first and only then a slower one at the cell membrane (Fig. 3.3c). The difference in the kinetics is due to the fact that C_{pip} is smaller than C_m. The current-dependent voltage drop at R_s naturally distorts the measured value of the membrane potential. This effect is compensated by generating a similar signal through a variable resistor and subtracting it from the potential generated by the pipette (Wheatstone bridge or bridge balance, Sect. 3.1.2).

▶ **Important:** During prolonged whole-cell recordings, resealing often occurs, i.e. the series resistance increases due to partial closure of the cell membrane, until, in extreme cases, a cell-attached configuration may be reached again. The compensation must therefore be checked and adjusted anew during the experiment. How to obtain the smallest possible series resistance and what to do if R_s increases during the experiment is discussed in Sect. 5.2.4.1.

Some amplifiers offer a so-called supercharging function for voltage clamp measurements to compensate for R_s (Armstrong & Chow 1987). In this case, the command voltage is also increased over the set value, but not by the feedback circuit described above. Instead, an optimized fast compensation is calculated from the values of C_m and R_s using a given algorithm. This method does not lead to increased noise or to oscillations, and very efficient compensations can be achieved for changes in the command potential. The method is particularly suitable for the study of voltage-activated conductances, but it does not improve the measurement accuracy at constant voltage. Cellular (membrane) currents that are slowed down by the series resistance are also not corrected.

Discontinuously operating amplifiers (dSEVC) eliminate the errors due to R_s and thus do not need this compensation, but the duty cycle must be set precisely.

3.2.2.2 Rules for Estimating Series and Membrane Resistance (R_s and R_m)

Finally, we would like to mention some quantitative rules, whose application provides important information about the cell and the quality of a whole-cell recording. The equations can be easily understood from Fig. 3.4 and are intended to deepen the understanding of the basic principles of electrophysiological measurements. Many modern patch clamp amplifiers and their associated software automatically measure and compensate for R_s, C_{pip} and C_m and provide the corresponding values, so that they do not have to be calculated in everyday life. However, this automatic compensation is not always accurate and must be checked with test pulses. It does not hurt to recapitulate the origin of these electrical properties.

We only consider the case where the series resistance is small compared to the membrane resistance. If this is not the case (and this may happen), we have to admit that a solid voltage clamp measurement is not possible. This does not apply to current clamp recordings, which provide important information also in these cases, as long as the compensation of the pipette capacitance is correct (see above). To estimate R_s and R_m we again start from a square wave command voltage pulse (test pulse) in voltage clamp mode. The current response must not be filtered at too low frequency, as this will affect the rate of the rise and thus reduce the maximal amplitude of the measured current (3 kHz as the cut-off frequency is a reasonable value; Sect. 5.3). The sampling rate must also be sufficiently high, taking into account the filter settings (Sect. 5.3.2). The compensation of the

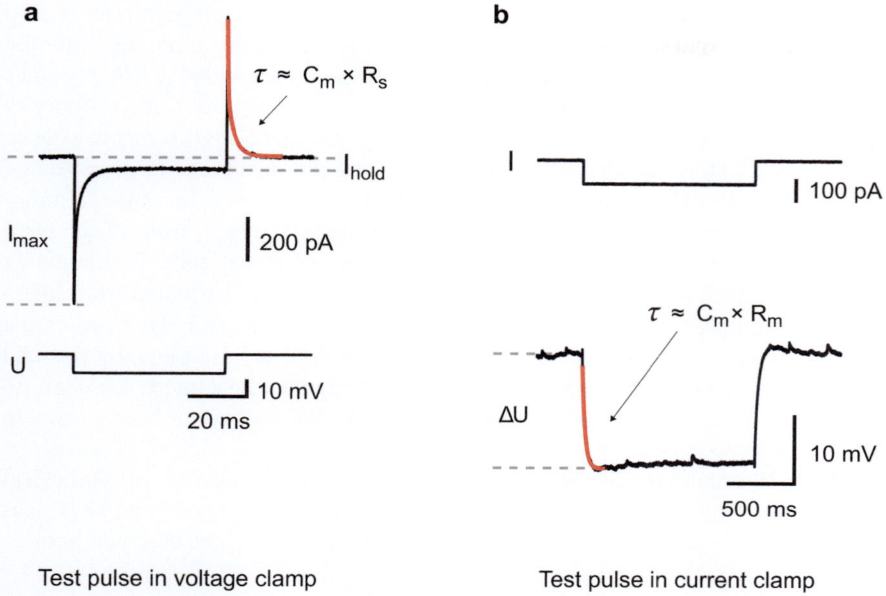

Fig. 3.4 Test pulses in the voltage clamp (**a**) and in the current clamp mode (**b**) for estimating capacitance, series resistance and membrane resistance. The red lines each show an exponential curve with a known time constant adapted to the signal (fitting)

membrane capacitance must be switched off for these measurements, as well as the compensation of the series resistance.

From Ohm's law and the definition of capacitance, the following estimates result for $R_s \ll R_m$:

$$R_m \approx \Delta U / I_{hold} \tag{3.1}$$

$$R_s \approx \Delta U / I_{max} \tag{3.2}$$

$$\tau \approx C_m \times R_s$$

Equation 3.1 is nothing more than Ohm's law, with R_s omitted (otherwise we would have to use $R_m + R_s$ instead of R_m). The resistance is calculated from the current required to maintain the new (10 mV more negative) voltage (I_{hold}) after recharging.

Equation 3.2 is also Ohm's law, but now for the beginning of the pulse, when the recharging begins. Here, only the series resistance is relevant, which limits the capacitive charge current. The equation implies: The smaller R_s, the greater the initial current (I_{max}) and vice versa. Note that the maximal current can only be reliably determined if the signal is not filtered at too low frequency!

Equation 3.3 gives the time required for the transient charging current to drop to 1/e (about 1/3) of its initial amplitude (I_{max}). The shorter this time, the faster the desired voltage in the cell will be reached. The calculation with $C_m \times R_s$ becomes qualitatively understandable if one realizes, that a large capacitor takes longer to recharge than a small one (τ proportional to C_m) and that a high series resistance ("bottleneck") slows down the current for charging the membrane (τ proportional to R_s). This approximation is valid only for $R_s \ll R_m$ and only under the assumption that the capacitance of the membrane is located directly in front of the electrode (usually at the soma of a small, more or less spherical cell). In complexly shaped cells such as neurons, the membrane capacitance is distributed over different compartments (soma, main dendrite, smaller dendrites etc.). This makes the charging process much more complex and voltage control is not equally fast and complete in all cell compartments. The total membrane capacitance can then no longer be estimated as simply as described above. We will come back to this in Sect. 5.2.4.

Electrical conditions of the recording configuration can also be determined in current clamp mode (Fig. 3.4b). Here, the amplitude of the current pulse (I_{inj}) is preset and the following equations apply for the recharging of the cell membrane:

$$R_m = \Delta U / I_{inj}$$

$$\tau = C_m \times R_m$$

The series resistance can not be derived from the measurements shown in Fig. 3.4b because it has already been compensated for in order to obtain a meaningful measurement, as described in Sect. 3.2.2 and Fig. 3.3.

Although most readers will probably not bother with the technical details described here, we recommend documenting the capacitance, series resistance and membrane resistance of each recorded cell (and to do this several times for long measurements). You should also include a simple hyperpolarizing test pulse into your recording protocols in order to be able to determine these values retrospectively for each measurement. This is extremely helpful for later questions about the validity of data! In larger series of experiments, it is useful to look for correlations between the above parameters (R_s, R_m, τ) and the recorded data. This is especially important to detect distorting effects of the series resistance-if the data depend on R_s, this is a sure sign of an experimentally induced bias.

3.2.3 Offset-Potentials

Offset potentials are voltages that do not originate from the biological preparation, but from transitions within the recording circuit: silver wire—pipette solution—bath solution—grounding electrode. Offset potentials arise between different electrolyte solutions or between electrolyte solutions and metal electrodes. They add up to the potential of the membrane and may cause significant errors, especially if the bath solution is changed during the experiment. If you want to characterize voltage-dependent processes, you need accurate values of U_m and must prevent offsets or at least know about them and compensate for them. The following sections describe the major sources of offset potentials.

3.2.3.1 Electrode Potentials

Chloridized silver wires or silver pellets (Sect. 4.4.4) are used at the interface between the measuring equipment and the pipette or bath solution. The coating with silver chloride (AgCl) allows the current to flow in both directions: If the silver wire is more negative than its surroundings, chloride ions enter into the solution, if it is more positive, chloride ions combine with silver to form AgCl. In the silver wire itself, electrons flow to generate Cl^- or to take up the negative charge of Cl^-—and this is how the signal from the ionic currents is transmitted to the measuring circuit (Fig. 3.5a).

Of course, the principle only works as long as there is no bare silver on one of the surfaces (usually the electrode wire). In this case, the bi-directional transition of chloride ions described above cannot take place, so that silver ions are dissolved and various cations or anions of the solution are deposited on the wire. These processes lead to increasing positive or negative polarization of the silver wire and to distortion of all measured potentials. It is therefore important to ensure that the electrode wires and grounds are well-chloridized. At the latest when unexplainable

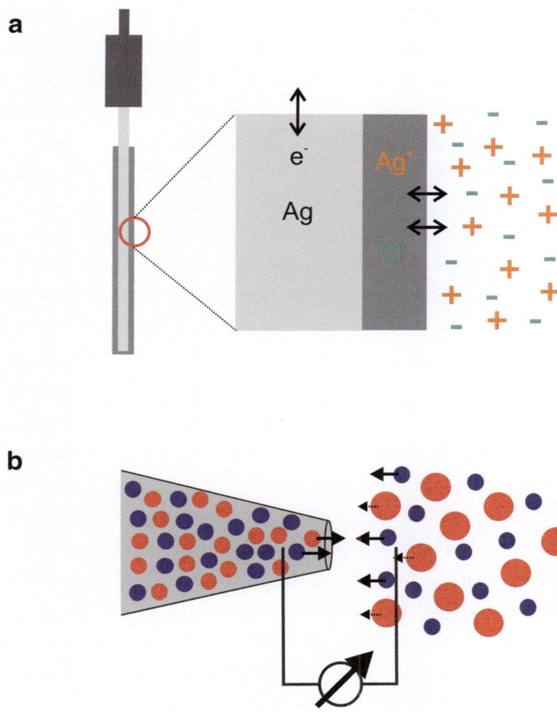

Fig. 3.5 Offset potentials. **a** Transition between silver wire (light gray), silver chloride layer (dark gray) and surrounding electrolyte solution (Sect. 3.2.3.1). **b** Development of a junction potential: The extracellular solution contains a particularly large ion that is difficult to move, all other ions are approximately equally mobile. The differences in mobility of the ions create an asymmetry that leads to a potential at the pipette opening (Sect. 3.2.3.2)

changes (drifts) of the measured potentials occur, the wires should be changed or re-chloridized.

Another systematic error can be caused by changes in the chloride concentration of solutions. The solubility product of AgCl is constant, with the proportion of silver being very low and the solubility of Cl^- being high. If the Cl^- concentration in the bath or electrode solution is increased or decreased, correspondingly more or less chloride ions will pass from the electrode into the solution. According to the Nernst equation, a tenfold change in Cl^- results in a potential of about 60 mV. If you switch between bath solutions with different chloride concentrations during the experiment, you should definitely work with an agar bridge, where the Ag–AgCl transition remains in a constant environment (Sect. 4.4.4).

3.2.3.2 Junction Potentials

At every junction between two solutions, there are diffusion processes of ions from one side to the other. This happens as soon as we dip the open pipette into the bath and the intracellular and extracellular solutions come into contact. If all anions and cations of the two solutions are equally mobile, the exchange has no effect on the potential. For example, the intracellular solution could mainly contain KCl, the extracellular NaCl—then everything is fine. But if one of the solutions contains a particularly large ion with low mobility, it will diffuse less easily into the other solution than the smaller ion present there. This creates a potential at the electrode tip that can easily exceed 10 mV (Fig. 3.5b).

Let's consider an intracellular solution containing potassium gluconate ($C_6H_{11}KO_7$). The gluconate ions are naturally less mobile than the small Cl^- of the bath solution. Therefore, more chloride diffuses from the bath into the electrode than gluconate diffuses out of the electrode into the bath. This creates a liquid junction potential (LJP) that is about 15 mV more negative in the pipette than in the bath. By convention, the LJP is given from the direction of the bath electrode, so +15 mV. However, as soon as we have formed a patch clamp configuration and thus closed the opening of the pipette with a membrane, the LJP disappears, because ions can no longer diffuse freely between the pipette opening and the bath. Usually, the current-free situation is set as V = 0 mV with the open pipette in the bath, i.e., the potential is "zeroed". In this case, the setting to "0 mV" on the amplifier is misleading—in fact, there is an additional -15 mV on the side of the pipette. After the seal is formed, the negative transition potential is now eliminated, so the input signal at the amplifier has become 15 mV more positive. A potential difference of 0 mV between the pipette and the bath electrode therefore corresponds to a value of +15 mV at the amplifier. If we now want to measure in the whole-cell mode at a membrane potential of -80 mV, for example, we would have to set a command potential of -65 mV to account for this offset (from +15 mV to -65 mV the voltage difference is then -80 mV). It may be advantageous to take the LJP into account when initially "zeroing" the open pipette or to enter it into the software. In this case, we would not set 0 mV, but the expected potential of +15 mV after the seal formation, so that all subsequent command potentials actually apply to the cell as membrane potential.

There are good open-source programs for calculating the LJP for different solutions. Some of the software packages for patch clamp measurements also include this option, so that you can avoid errors without much effort. You can also measure the LJP yourself by first dipping the patch pipette into a bath containing the pipette solution (junction potential equal to zero). The bath is then filled with the extracellular solution to be used later—now the actual expected junction potential is generated and can be read directly on the amplifier. Respective protocols have been published (Neher, 1992) and there are many step-by-step instructions on the internet.

References

Armstrong CM, Chow RH (1987) Supercharging: A method for improving patch-clamp performance. Biophys J 52:133–136

Dolzer J (2021) Patch clamp technology in the twenty-first century. Methods Mol Biol 2188:21–49

Moroni M, Servin-Vences MR, Fleischer R, Sanchez-Carranza O, Lewin GR (2018) Voltage gating of mechanosensitive piezo channels. Nat Commun 9:1096

Neher E (1992) Correction for liquid junction potentials in patch clamp experiments. Methods Enzymol 207:123–131

Sigworth F.J. (1995a) Design of the epc-9, a computer-controlled patch-clamp amplifier. 1. Hardware. J Neurosci Methods 56:ES/*ME2.

Sigworth FJ (1995b) Electronic design of the patch clamp. In: Sakmann B, Neher E (eds) Single-channel recording, 2nd edn. Springer, Boston, MA, pp 95–127

4 The Patch Clamp Rig and Technical Components

Often, work in electrophysiology begins with training on an existing rig—whether as part of an internship, an academic thesis, or a new job. Sometimes, however, you have to set up a new setup from scratch, which has its advantages, since you know it from the ground up. In any case, it is advisable to think systematically about which technical components are needed for the planned patch clamp recordings and what needs to be taken into account. If you have to (or are allowed to) build your own setup, you will definitely understand the essentials afterwards—according to the physicist and Nobel laureate Richard Feynman: "What I cannot create, I do not understand". So: don't be afraid of technology!

Although patch clamp rigs vary in detail depending on the type of experiment, a few key components are common to virtually all setups:

- **Optical components**: Usually, the sample must be made visible with a binocular or a microscope so that the target structure can be seen and approached with a pipette. Often, additional optical methods are used, such as fluorescence imaging, confocal microscopy or optogenetic stimulation techniques.
- **Recording chamber**: The sample must be stably positioned in a suitable chamber where it remains in good condition for the duration of the measurement (order of magnitude: hours). For this purpose, the chamber is usually perfused with a cell-friendly nutrient solution. Additional functions may have to be considered in the design of the chamber, e.g. the temperature control of the recording solution or rapid solution changes for pharmacological experiments.
- **Micromanipulator**: The patch clamp technique relies on a close contact between the pipette and the membrane. For positioning the pipette, micromanipulators are required that must be precise (µm) and very stable over a long period of time (hours). Tension- and vibration-free fixation of the manipulator is particularly important.

F. C. Roth et al., *Mastering the Patch Clamp Technique: A Practical Guide*, https://doi.org/10.1007/978-3-662-70745-6_4

- **Pipettes and pipette holders**: Making pipettes requires a good pipette puller and some experience to find the ideal shape for the particular recording. The pipette is fixed in a special holder, which must be mechanically stable and airtight in order to be able to deliver negative and positive pressure to the pipette.
- **Electronic components**: The core of the patch clamp setup is the amplifier for measuring electrical currents and potentials. Often, additional devices are added for recording other electrical signals (e.g. field potentials), for preprocessing of recorded data (e.g. filters against noise signals) or for electrical stimulation.
- **Data processing**: The recorded signals are usually digitized, displayed on a computer screen (less often also on an oscilloscope), and stored and analyzed on a PC. This requires analog-to-digital converters that transform output signals from amplifiers and other recording devices (analog voltage signals) into digital data. These converters are already integrated in some amplifiers and digital oscilloscopes. Allow plenty of time for documentation and later analysis of the data. Rule of thumb: 1 day of measurement = 1 to 2 days of analysis.

In this chapter, we will discuss the general components of patch clamp rigs, some of which are shown in Fig. 4.1. We will explain the basic functions of the instruments, introduce different types, and point out what to consider when buying and installing them. Often, it makes sense to build or adapt certain parts yourself—this is also covered in this chapter.

4.1 The Optics

For a patch clamp recording, you usually want to see the cells or cell groups (tissue structures) as you approach the target with the pipette. Sometimes, however, you may need to see down to subcellular structures, such as when recording from axons or dendrites. Almost always, you will need optical tools to make the sample and the pipette visible with the required resolution.

In modern physiology, optical methods play an enormously important role—this also applies to electrophysiology. Many patch clamp rigs contain either sophisticated microscopes (e.g., fluorescence imaging; confocal microscopy) or devices for additional optical methods (e.g., optogenetic stimulation, uncaging). In short: you can't do it without optics. Fortunately, not without electrophysiology either… For patching visualized cells, a light microscope with a total

Fig. 4.1 **a** The patch clamp rig. 1: PC system for data acquisition, 2: oscilloscope, 3: PC screen for microscope camera, 4: patch clamp amplifier and A/D converter, 5: manipulator control, 6: manipulator with headstage, 7: microscope camera, 8: microscope, 9: table with recording chamber, 10: Faraday cage, 11: vibration-damped table. **b** Schematic of the signal chain. ▶

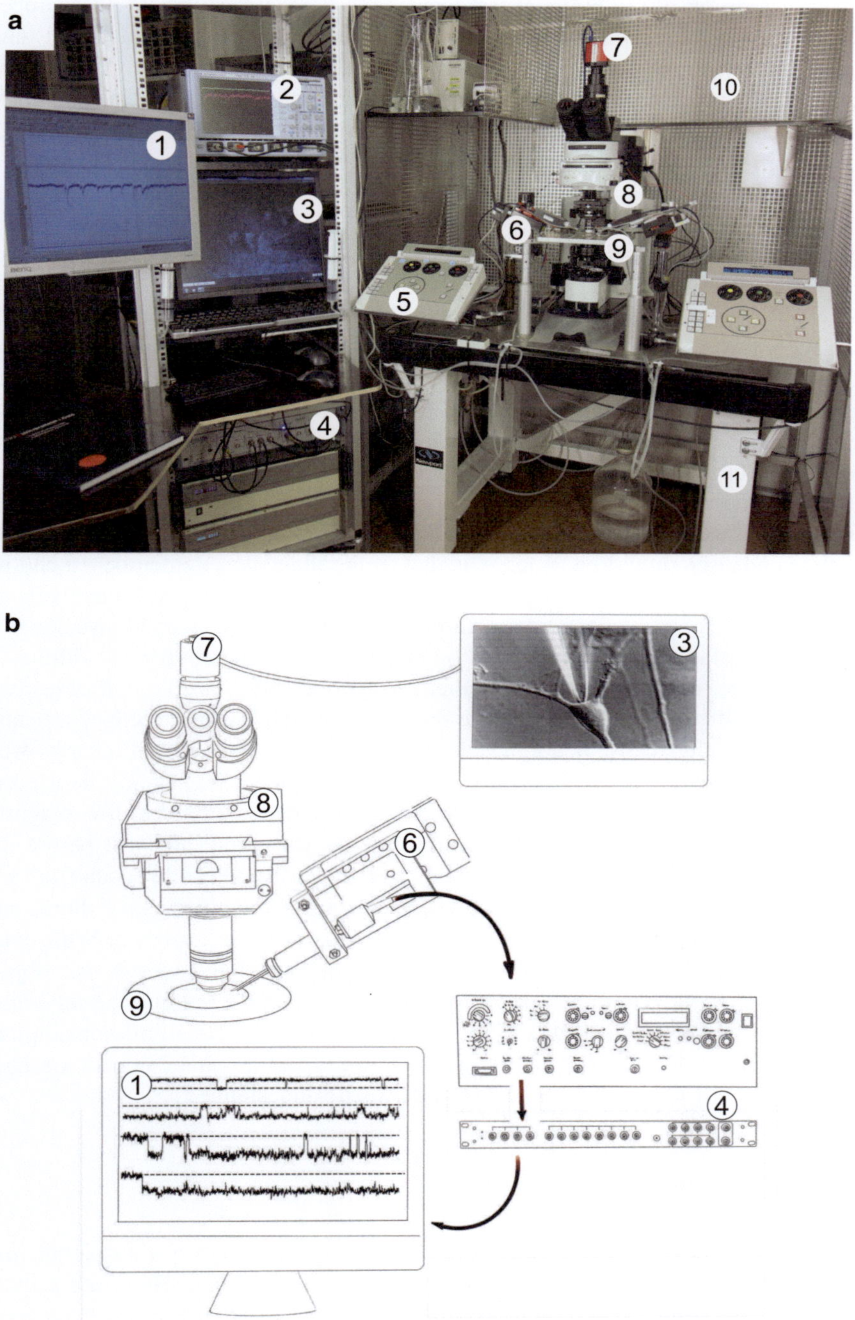
a
1
2
3
4
5
6
7
8
9
10
11
b
7
3
8
6
9
1
4

magnification of 200 to 600 × at the eyepiece is used, which should be equipped with contrast enhancement optics. For thin specimens (e.g., monolayer cell cultures), the objective is usually aimed at the cells from below (inverted microscope), leaving plenty of space above the preparation for the pipette. For thick, multilayered specimens (e.g., brain slices, organoids), you usually have to look at the specimen from above (upright microscope).

4.1.1 The Inverted Microscope

With an inverted microscope (Fig. 4.2a), the objective is pointing upwards, so you look at the cells from below, leaving plenty of space above for the recording pipette, the bath perfusion, and other components. The space above the specimen is only limited by the illuminator, so you must use condensers with a sufficiently large working distance. Because the objective of the inverted microscope can come very close to the bottom of the recording chamber, relatively high-magnification objectives with a large numerical aperture can be used; however, oil immersion is cumbersome to handle and usually not necessary. The numerical aperture of the condenser should be as large as that of the objective. Remember to choose a long-distance version, so that the working distance remains sufficiently large (e.g. > 5 cm). Contrast-enhancing methods are used to provide a good visualization of the cell surfaces. Usually, a phase contrast device is sufficient for cell cultures.

In practice, condensers with a numerical aperture of about 0.6 and objectives with a magnification of 20 to 40 × are used, with which "typical" cells (for example neurons, glial cells, heart muscle cells or fibroblasts) are sufficiently well visible. In addition, a lower magnification objective (4 to 10x) should be used to locate the cells in an overview. The pipette is also easier to find at low magnification, and you will usually switch to higher magnification only just before the cell is approached. Initially, most people have some difficulties to focus from the bottom through the sample. This makes the search for the pipette tip tedious, and some glass breakage at the bottom of the chamber is to be expected. With practice, the procedure quickly becomes routine. On most microscopes, the objective revolver is moved up and down for focusing. The table holding the recording chamber with the cells is static. If it is sufficiently stable, a small micromanipulator for the recording pipette can be mounted on top of it. This compact arrangement avoids unwanted movements between pipette and sample.

4.1.2 The Upright Microscope

When working with "thick" specimens that are not sufficiently transparent, use an upright microscope (Fig. 4.2b) rather than an inverted one. Here, the sample is illuminated from below and viewed from above, i.e., from the same direction as the pipette. Upright microscopes can be used to record from visually identified cells in tissue sections or other three-dimensional specimens (e.g., organoids).

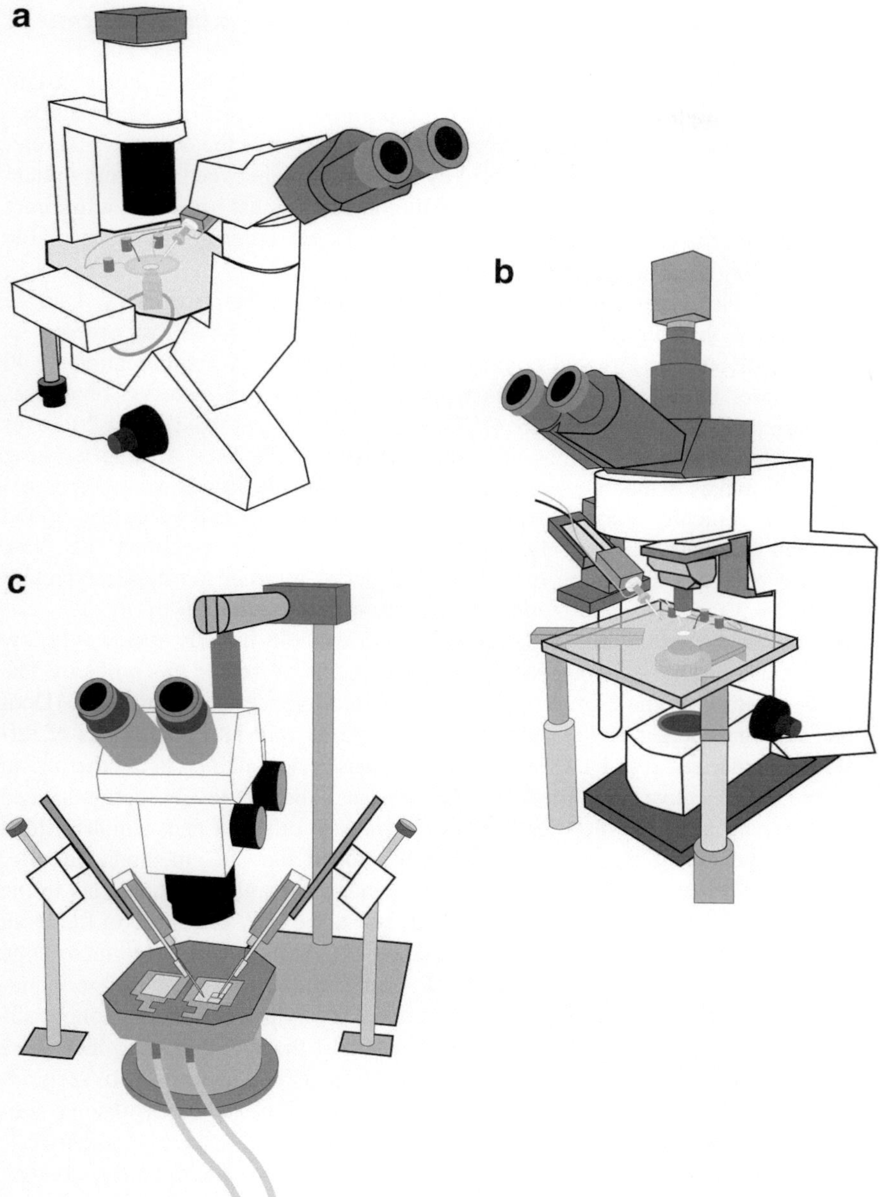

Fig. 4.2 Inverted (**a**) and upright (**b**) microscope for patch clamp experiments. **c** Dissecting microscope above an interface chamber

Working on thick specimens with an upright objective poses three problems:

1. The interface between aqueous extracellular media and air is not constant and distorts high-resolution images. This problem is solved by water immersion (WI) objectives, where the front lens is immersed in the aqueous medium. This eliminates any distortion caused by the transition between different optical media. Of course, the metallic parts of the objective must not come into direct contact with the bath. Isolated WI objectives for electrophysiology are available from all major manufacturers.
2. The objective comes close to the specimen, so the remaining working distance is small. With today's common (not quite cheap) objectives, this problem is largely solved. For targeted control of individual cells, a magnification of 40 to 63 times is recommended. The working distance of suitable WI objectives is given as the working distance (WD) in mm and ranges from about 2 to over 5 mm, depending on the magnification and type of the lens. A large working distance makes it much easier to insert the pipette! For this reason, the objectives are conically shaped at the tip so that the pipette can be easily guided through the narrow gap between the objective and the specimen. For this, it must be brought into the recording chamber at a rather flat angle, typically about 25–30° from the horizontal plane (Fig. 4.2b, left).
3. Due to light scattering in the sample, the individual cells initially appear very low in contrast. Several methods have been developed to overcome this problem. The most common is the Nomarski Differential Interference Contrast (DIC) (Dodt and Zieglgansberger 1990). Here, polarized light is split into many pairs of differently polarized rays by a prism in the condenser, recombined in another prism above the objective, and filtered in the analyzer. The contrast is then enhanced by the interference of the slightly shifted rays with different path lengths. Many microscope manufacturers have further developed the DIC method and offer slightly different solutions. Contrast enhancement by optical components in the beam path of the illumination has also proven to be very effective. In the Dodt Gradient Contrast (Dodt et al. 1998, 2002), the light between the microscope lamp and the condenser is first filtered through a circular segment (quarter annulus) in an additional tube and then spread out as a brightness gradient using a diffuser. This creates a gradient in the illumination of the specimen, in which there are also interferences between waves with different path lengths. Both methods efficiently enhance the contrast. The oblique contrast illumination works in a similar way, where the light is directed obliquely into the specimen and laterally offset through the condenser. The gradient methods are somewhat more flexible than DIC and can be more easily combined with fluorescence microscopy.

For thick specimens (e.g., brain slices), infrared (IR) illumination has become a standard, using light with a particularly long wavelength (775–900 nm) that scatters less in the tissue. This method requires only an IR filter between the light source and the specimen (available from manufacturers of special glasses or microscopes) and an IR-sensitive camera.

▶ Sometimes you can find inexpensive video cameras that have a built-in infrared blocking filter that can be easily removed.

Finally, confocal techniques ranging from the "simple" confocal to 2-photon microscopy are ideal for thick specimens. However, these solutions are quite expensive and usually only necessary for special applications. This is the case, for example, when high-resolution fluorescence signals in deeper structures are studied or when other methods such as subcellular live imaging or highly focused stimulation techniques (optogenetics or photolysis of caged compounds) are used.

Many manufacturers also offer modules that magnify the image in the beam path of the microscope. This allows working continuously with a medium-magnification objective (e.g., 25x; often in combination with fluorescence imaging) or to record from very small structures with a high-magnification lens. However, the intermediate magnification only optically "inflates" the image without increasing the resolution. We see this variant as an interesting option, but by no means a "must have".

If you need to shift the field of view in your experiment, for example to position multiple recording or stimulation electrodes, you should mount the microscope on a stage that can be moved in two axes (X-Y stage) so that it can be shifted relative to the recording chamber. The recording chamber and the micromanipulator must be mounted separately, i.e., independently of the microscope. Such stages are offered by specialized equipment manufacturers with manual micromanipulators or with motor control for automated positioning. Alternatively, the recording chamber and manipulators can be moved together on a platform if the microscope must be fixed to immovable parts of the beam path. In practice, the result is the same.

There are many ways to set up an upright microscope for patch clamp experiments. You should be sure about the conditions under which you want to record: How many recording, stimulation or application pipettes do I need? Do I need a fluorescence device and therefore a special objective? Are there special requirements for the recording chamber? It is always worth asking experienced colleagues and taking a look at their recording equipment. Many microscope manufacturers and companies specialized in electrophysiology have great expertise in the requirements for patch clamp recordings and can provide competent information. Many also offer complete solutions where the microscope is purpose-built and optimized for patch clamp experiments.

4.1.3 Stereo Microscopes

In some cases, it is not possible or necessary to record from a specific, visually identified cell. This is the case, for example, when the optical conditions are too poor (e.g., when recording from deep structures in vivo) or when the specimen is placed in a chamber that is not suitable for high-resolution microscopy (e.g. an interface chamber; Sect. 4.2.3). In such cases, a low magnification view of the specimen can be used to insert the pipette into the tissue at the correct location and angle. The actual patch clamp configuration is then achieved solely by monitoring the response to an electrical test pulse, which changes characteristically as the cell is approached and the seal is formed (Sect. 5.1). The procedure is similar to blind flying and is therefore called blind patch.

For this purpose, suitable stereo (inexpensive) or surgical (expensive) microscopes are available (Fig. 4.2c). The magnification can be changed step by step or by zooming, typically up to about 100x. At higher magnifications the depth of field becomes too small. Of course, you should look for a large working distance to provide space for pipettes and manipulators—but for most microscopes this is about 10 cm or more, so the layout is rather luxurious compared to high-resolution microscopes.

A separate, not too weak lamp is useful, preferably with a flexible optic fiber, so that the power supply and cable remain at some distance from the specimen. The standard light sources supplied by the microscope manufacturers are often too weak and sometimes cause interference signals due to electromagnetic induction. One trick is to remove the lamp's AC power supply and instead use a DC power supply placed outside the Faraday cage (Sect. 4.2).

4.1.4 Cameras

In most cases, the specimen is not viewed through eyepieces, but the image is made visible on a PC or separate monitor via a camera. Some manufacturers even do away with eyepieces altogether. With a comfortably visible camera image, you not only avoid neck and back pain, but often have a much better image with stronger magnification (~2x). By using a microscope camera, lenses with the largest possible numerical aperture can also be used, which improves the light output and resolution. The associated loss of contrast can be compensated by digital post-processing of the image (contrast enhancement, averaging or background subtraction). When using infrared (invisible) light, a camera system is of course mandatory.

A microscope camera has a further advantage: While establishing a seal, the experimenter can simultaneously observe the pipette and the electrophysiological signals on the monitor (Sect. 5.1). Even during the actual patch clamp measurement, the cell under investigation can be continuously observed and additional electrodes can be positioned without having to look directly into the microscope—this helps avoiding vibrations and electrical interference. We will discuss special microscopy techniques in Chap. 6.

► In the case of sensitive or complex samples, poor visibility of cells is sometimes a sign of poor sample "quality". The general term "sample quality" is not easy to define. Ultimately, it refers to damage of the cellular integrity or homeostasis (pH, ions, osmolarity, O_2 etc.) that has occurred during exposure, cutting, transport or storage of the specimen. It is often worth the effort to optimize tissue quality and to consult colleagues who have experience with this particular specimen.

4.2 Microscope Stage and Mechanical Components

4.2.1 Optical Table, Microscope Stage and Cage

Even the slightest movement between the sample and the pipette will disrupt the highly sensitive patch clamp recordings. Therefore, the setup must be shielded from vibration and any mechanical impact from the environment. Every building causes such interferences, whether from electrical machines, passing traffic, or simply from our dear colleagues slamming the door.

4.2.1.1 The Anti-Vibration Table

A vibration isolation table reliably eliminates many mechanical disturbances. Typically, these tables consist of a stable metal base and a 100–300 kg heavy, solid or internally braced table top. The resonance frequency of such tables is very low (a few Hz), which means that faster movements are efficiently damped (e.g., shocks from slamming doors, from refrigerator compressors, etc.; on the other hand, building vibrations on high floors are often low-frequency and are not well absorbed). The table top is usually mounted on pressurized shock absorbers, built into the table legs. Therefore, a compressed air supply should ideally be available in a patch clamp lab. A less elegant alternative is to use small compressors or well-secured(!) bottles with compressed air. Tables of various sizes and designs are available from specialized laboratory suppliers. Often, they are offered as ‘optical tables’, which is somewhat discriminatory against us electrophysiologists.

When **choosing an anti-vibration table**, one should consider the required size and stability. For simple setups consisting of a microscope, micromanipulator(s), and recording chamber, about 100 × 80 cm is sufficient, but for larger systems (e.g., with laser light sources placed on the same table), 200 cm edge length may be necessary. All common anti-vibration tables absorb vertical vibrations (with price and quality differences, of course), but not all tables are well protected against horizontal impacts. Where these occur, models with internally braced hollow metal supports are particularly suitable. Only in extreme cases are tables needed that measure movements and compensate for them with active, piezoelectrically generated counter-movements. If no anti-vibration tables are already present in other labs nearby, a test should be arranged with a manufacturer or distributor before deciding on a model.

One convenient feature of the table top are predrilled threaded holes. Of course, if you have good tools or a workshop, you can cut these yourself as needed. If you use stone tabletops (very heavy, very robust and relatively inexpensive), you may want to add a metal plate on top with holes for mounting equipment. Magnets are also very useful as holders—the metal surface of the table should thus be magnetizable.

> If you work in a very stable building, you may not need to buy a table. For a homemade one, you place a heavy stone or metal plate on vibration-absorbing objects such as inner tubes from children's scooters, tennis balls or foamed materials on a sturdy table. So-called weighing tables for laboratory scales may also be suitable and are often found as abandoned objects in the institute's basements. However, homemade tables often have unwanted resonance frequencies and are not of the same quality as commercial ones. If possible, we recommend purchasing a professional table.

When **positioning the setup,** it is important to note that there are large differences in mechanical stability between different buildings, rooms, and locations within each room. Therefore, sufficient time should be invested in finding the optimal location. To do this, testing is more effective than applying fixed rules. Nevertheless, here is some general guidance: It is best to place the table in a corner of the room or near a load-bearing wall, where vibrations from the floor are expected to be lowest. The table should not be too close to doors and passageways to avoid unnecessary conflicts with colleagues. Proximity to refrigerators and freezers is also unfavorable, as their compressors typically start up during the crucial experiment, which may cause significant interference signals. Often overlooked sources of disturbances are room ventilation and air conditioning units, which can cause variable and extremely persistent movements of the pipette. If the system cannot be avoided or turned off, it may be necessary to shield the setup with curtains, foils or metal plates.

To avoid accidental contact with the table during the experiment, it is recommended to surround the vibration table with a kind of solid "over-table". This over-table resembles a matryoshka doll, being higher than the actual table top and leaving a cutout for the microscope, the stage with recording chamber and the manipulators. The Faraday cage for electrical shielding can be placed on this "over-table" (Sect. 4.2.1.3). Many manufacturers offer Faraday cages that can be directly mounted to the rigid base frame rails around the table top.

Only essential components should be placed on the anti-vibration table: microscope, recording chamber (on a stage), micromanipulator with pipette holder and preamplifier. Ideally, all items contacting the cells (directly or indirectly) should be mechanically coupled to prevent relative movements. Cables and tubing should be neatly laid out and kept as short as possible. They should not form loops and should be tied to solid mechanical components to avoid movements.

4.2.1.2 Configuration of the Setup

In the immediate vicinity of the sample there are usually four components: the microscope, a recording chamber for storing the specimen, the micromanipulator with pipette holder and pipette, and the preamplifier (the heart of the patch clamp amplifier, more on this in Sect. 4.3.1). Often, the smaller objects are placed on a separate platform, called a microscope stage, which in turn rests on the much larger anti-vibration table. There are many different configurations of electrophysiology rigs—here we present four basic variants:

1. **Compact arrangement on the microscope stage:** Most microscopes have their own stage on which to place the specimen. If this stage is sufficiently large and stable, the recording chamber and the micromanipulator can be attached to it, often also the preamplifier. This results in a very compact setup that minimizes relative movement of the components. Small, lightweight micromanipulators should be used, which are available from several companies. A typical application are "simple" single-electrode recordings from cultured cells, using an inverted microscope.
2. **Separate stage:** In the slightly larger and more flexible variant, the mechanical components are separated from the microscope. Typically, they are then mounted on a separate stage, whose height and dimensions are matched to the microscope. This stage is fixed to sturdy columns that rest on the large anti-vibration table. It carries the recording chamber and the micromanipulator(s), so that no relative movements can occur. Alternatively, the manipulators can also be mounted independently of the recording chamber on tripods or posts—in this case, all elements can be connected to each other by horizontal rails to avoid relative movements. Such posts (optical rails) together with the necessary clamps are available from various manufacturers. In this arrangement, the microscope, which is decoupled from the actual recording chamber, can be mounted on a separate X-Y stage so that it can be moved relative to the specimen. Typical applications are recordings from brain slice preparations, recordings with multiple electrodes (multiple patch clamp recordings or combined patch clamp and extracellular recordings), or the use of additional components close to the preparation (electrical stimulators, devices for rapid substance application, optical fibers, etc.).
3. **Blind-patch setup:** In some applications, it is not possible or necessary to see the individual cells and the pipette tip. In these "blind" patch applications, the experimenter is guided by coarse anatomical structures such as cell layers or organ boundaries and establishes the patch clamp configuration solely based on the electrical signals (Sect. 5.1). In such cases, the high-resolution microscope can be replaced with a simple stereomicroscope with large working distance. The recording chamber and manipulators can now be arranged much more freely. For example, the sample can be placed in a temperature-controlled and gassed interface chamber (Sect. 4.2.3), and the pipette can be inserted into the specimen at a very steep angle. Typical applications are recordings from brain slices, often in combination with other electrophysiological or optical techniques such as "sharp" intracellular recordings, extracellular recordings or large-scale recordings of network activity with calcium or voltage-sensitive dyes.
4. **In-vivo patch clamp:** More recently, patch clamp measurements have been increasingly performed on living animals (mostly rodents), sometimes even

during active behavior in the awake state. The setup of such recordings is fundamentally different from in-vitro recordings. In vivo recordings are usually performed as “blind” patch clamp recordings in the whole-cell configuration (Sect. 6.8). Usually, only a dissecting microscope is placed above the animal to roughly position the pipette, which is inserted through a small hole in the skull and dura. It is essential that the animal is stably positioned and stress-free, so that there is no relative movement to the pipette. For anesthetized animals, a stereotactic frame is used, which can be purchased for mice or rats. Alternatively, a holder can be attached to the head to secure the animal during the recording. Mice or rats can even be trained in this head-fixed configuration, so that they can move on a treadmill or ball without being stressed, such that electrophysiological signals can be studied during active behavior. Such experiments are carried out in specialized laboratories and require very specific expertise, such that we will not describe them in detail here. Some patch clamp recordings have even been made in freely moving animals, but this is certainly not the subject of an introductory book.

4.2.1.3 Electrical Shielding

A characteristic of the patch clamp technique is its extremely large amplification. The tiny electrical signals from individual cells or ion channels can easily be masked by electrical background noise—so the signal-to-noise ratio must be optimized. The main source of low-frequency interference is the so-called mains hum, i.e., the interference of the electromagnetic alternating field, which is generated by the public power line and transmitted into the recording setup via metallic components and cables (frequency 50 Hz in Europe or 60 Hz on the American continent and in some other countries). In addition, there is high-frequency noise, which is caused by various processes which we summarize in Sect. 5.3.1. There, we also give detailed advice on how to reduce the electrical noise and hum. However, a few simple rules should be followed when setting up the rig to avoid frustrations later on.

The introduction of mains hum can be reduced by using a Faraday cage to shield the inside of the setup from electromagnetic fields. Such a cage is not always required—in fact, it can be omitted in many cases. However, in other cases it can provide significant relief, and we strongly recommend it when in doubt. Do-it-yourself solutions such as aluminum-wrapped cardboard boxes near the recording chamber may have their charm, but efficient work with solid constructions is much more fun in the long run.

The cage forms a large hood over the entire recording setup. There are commercially available Faraday cages, but anyone with access to a workshop can easily build (or have built) a custom cage from metal profiles and sheets or conductive foils with a metallic coating. Some use so-called μ-metals with high magnetic permeability for shielding, which shield magnetic fields slightly better than conventional metals. Small accessories, such as perfusion systems, can be conveniently magnet-attached to solid cages made of magnetizable metal. Modifications are particularly easy when the cage walls are flexibly attached to the frame with magnetic strips or with quick-release wing nuts.

In the case of strong electromagnetic disturbances, the rig should also be shielded at the front. This can be done with a hinged door or a conductive curtain that can be opened to change pipettes or specimens. The mechanical decoupling of the cage from the table top is important. The cage is either part of the above-described frame or it stands on its own legs. Never place the cage directly on the anti-vibration table!

Inside the cage (a few centimeters above the anti-vibration table top), a kind of windowsill can be installed on which small accessories can be placed and used as a support while working. After some time, a collection of various useful items such as tweezers, plugs, tools, unused cables, etc. typically accumulates there. However, electrical devices, especially those with AC connections, should be kept to an absolute minimum inside the cage, as each device can be an additional source of interference.

> **Caution:** All parts of the cage must be electrically connected and grounded. We provide instructions and tips in Sect. 5.3, but we do not want to hide the fact that grounding is a rather para-scientific matter. Some rigs, for example, suddenly start humming after a weekend, without having been noticeably modified. (Some bosses use this as a reason for asking you to use your rig in the weekends, too.)

4.2.2 Micromanipulator and Pipette Holder

4.2.2.1 The Micromanipulator

A micromanipulator is required to precisely position the patch pipette onto the usually tiny cell. There are a number of suppliers for micromanipulators that operate on different principles: mechanically, hydraulically or electrically, the latter being either motor-driven or piezo-driven. The hydraulic and electric micromanipulators are operated from a spatially separate unit that is connected to the actual manipulator by hoses or cables. These connections must be carefully fixed and strain-relieved (cable ties, adhesive tape, Velcro), so that no unwanted movements are transmitted to the pipette!

There are excellent brands of micromanipulators for all of the applications mentioned above, and in principle many types of manipulators are suitable for patch clamp recordings. On the other hand, a drifting, wobbling or humming micromanipulator will make any successful work impossible. We do not want to recommend any manufacturer—you should rely on proven equipment and ask your colleagues which types are suitable for your application. There are big price differences, but the micromanipulator is such an important piece of equipment that you don't want to cut corners in the wrong place!

What should you look for? A good micromanipulator must be compact, so there should be no long, thin levers between the base of the device and the pipette (and no long bars for mounting the pipette holder!). The size of the manipulator becomes more important the less space is available near the recording chamber, especially for multiple parallel recordings. If you want to place several micromanipulators in a confined area, you should seek the advice of experienced colleagues and contact specialized laboratory equipment companies. On the other hand, if you have a lot of space, it is a good idea to use a rather massive manipulator that guarantees high stability due to its own weight and the strong materials used.

A guaranteed movement accuracy of less than 1 μm is required (optimal is less than 100 nm), and the total movement ranges should not be less than 5 mm in all directions. In some models, this is achieved by a combination of drives for coarse and fine movements, often allowing several centimeters of adjustment. As a minimum requirement, there must be at least one fine drive that allows precise movements in the direction of the pipette axis. This allows the final approach to the cell when establishing the recording.

For quick and easy pipette changes, electrically operated manipulators usually have a 'home' function that allows them to move quickly back to the starting (home) position. Other models have joints or are mounted on rails that allow the manipulator to be easily moved away from the sample for convenient pipette changes. Mechanical stops allow the manipulator to quickly return to its working position for the next recording.

Mounting of the Micromanipulator

The micromanipulator should be mounted in such a way that one-dimensional movement in the direction of the pipette axis is possible when the pipette is close to the cell. However, with some practice, it is possible to deal with differently oriented axes, which then require an iterative approach process. As mentioned above, an important prerequisite for stable recordings is that there are no forces acting on the manipulator through tubing and cables. After all, the manipulator is moved quite a bit during each pipette change, and if there is any tension on the connection between the manipulator and the control unit, no stable recording will be possible. Therefore, a very careful strain relief near the manipulator is important. This can be done in various ways, from do-it-yourself solutions with tape (not recommended!) to cable ties to precision-machined clamps.

4.2.2.2 The Pipette Holder

For patch clamp recordings, the glass pipette must be fixed very stably, and it must be possible to apply positive or negative pressures inside the pipette. This is done by the pipette holder, in which the pipette is clamped tightly and airtightly by means of rubber seals (Hamill et al. 1981). A connecting nozzle leads from the inside to the outside to allow pressure changes through a connected tube. The pipette holder is usually attached directly to the preamplifier to avoid additional noise and mechanical instabilities caused by cables and long rails. Pipette holders are usually supplied with the amplifier and can be ordered in many variations.

Since dimensions and materials are based on decades of experience of the manufacturers, we advise against self-made designs, unless very special prototypes are required. Most pipette holders are made of polycarbonate, and Teflon (PTFE, polytetrafluoroethylene) is used for the insulating parts. Particularly (thermally) stable are pipette holders made of quartz. In any case, it is important that the pipette has no mechanical play during pressure changes. This is achieved, for example, by two separate clamping points of the pipette within the pipette holder or, for particularly high demands, by additional fixation of the pipette directly to the manipulator. If mechanical drift occurs during use, it is often sufficient to replace the rubber seal within the pipette holder.

▶ The inner diameter of the holder must match the outer diameter of the pipette in any given case! All manufacturers offer holders for common pipette dimensions. A separate electrical shielding of the holder increases the noise and should not be used without good reasons!

A tube is connected to the side port of the pipette holder (Fig. 4.3), through which suction or positive pressure can be applied to the inside of the pipette (Sect. 5.1). For this purpose, a cut-off pipette tip is attached to the end of the tubing as a mouthpiece or a standard connector for a syringe. It is recommended to install a valve in the tubing (for example, a standard three-way valve from infusion sets) that can be closed to maintain the applied pressure. In addition, a manometer can be attached to the tubing in order to develop a sense of the pressure and the tightness of the system. The relevant range is about ± 200 cm H_2O (200 mbar, 20 kPa, 2.9 PSI). Inexpensive manometers are available from electronics or laboratory

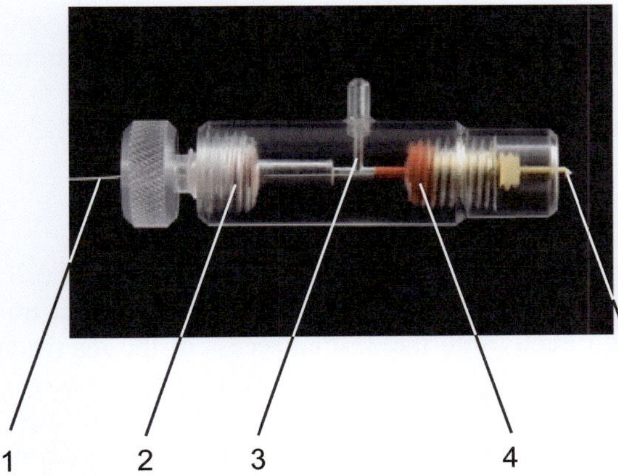

Fig. 4.3 Pipette holder. 1: silver wire; 2: front seal for fixing the pipette; 3: connector for tube for pressure application; 4: rear seal; 5: connector to preamplifier (Image courtesy of ALA Scientific Instruments, USA/npi electronic, Tamm, Germany)

supply stores. Extreme accuracy is not necessary for most applications, and with practice, you will soon get a feel for the right pressures.

The pressure tubing provides a direct mechanical connection to the pipette holder. This bares the risk of unintentional movements being transmitted to the pipette, which will disturb the stability of the recording. Therefore, the tubing should be very carefully tension-relieved near the preamplifier by fixing it to the micromanipulator, to the posts, or other suitable objects. We use rather soft, flexible tubes, but others swear by hard materials, e.g. Teflon. There seems to be no absolute right or wrong here.

▶ **Caution:** The pipette holder must always be dry and clean. For extremely noise-sensitive experiments it should be washed with isopropanol (not aggressive methanol or expensive ethanol) and dried well (e.g. with compressed air) before each experiment.

The electrical connection between the pipette and the preamplifier is made via a silver/silver chloride wire. This ends in a plug (matching the input of the preamplifier), where it is fixed either mechanically or by soldering.

▶ **Attention:** Solder residues can cause leaks at the back of the pipette holder, so that the pressure may not remain constant!

4.2.3 Recording Chambers

The ideal type of recording chamber depends mainly on the specimen and the experimental design. There are some basic types and several variations. They are available from specialized suppliers, but can also be made by a mechanical workshop. Plexiglass or Teflon (PTFE, polytetrafluoroethylene) are suitable materials, but biocompatible materials are now also available for 3D printing. The stages supplied with the microscope are available with cut-outs to fit chambers with standard dimensions fit (e.g. for a 35 mm cell culture dish). The major types are described below:

4.2.3.1 Simple Chambers

The simplest chamber for working with acutely isolated cells or cell cultures is the culture dish itself. Commercially available Petri dishes made of transparent polystyrene are inserted directly into the 35-mm recess of the microscope stage. The optical properties of these Petri dishes are moderate, but sufficient for high-contrast and not too flat cells. However, the working distance of high-magnification objectives is so short that, with inverted microscopes, the bottom of the Petri dish may already be too thick to see the pipette tip coming from above—so a weaker objective with a longer working distance may be more useful. In many cases, the cells are cultured on thin, round coverslips (1 cm in diameter, possibly coated with poly-D-lysine). These can be easily transferred into a recording chamber. If you do not have a special recording chamber with a milled recess for the cover glass, you can attach the cover slip with a drop of Vaseline to the bottom of a 35-mm culture dish.

Simple round recording chambers have two major limitations: They are difficult to perfuse completely and are not well temperature controllable (Sect. 4.2.4). Often, you want to wash out the solution in the chamber and replace it with another bath solution (e.g., a solution containing a drug). However, Petri dishes have very unfavorable conditions for a uniform flow (this can be easily demonstrated by washing in dyes such as trypan blue). Therefore, we recommend specially made recording chambers with a smaller inner volume and better flow conditions. Such chambers can be custom built or purchased from various manufacturers.

4.2.3.2 Temperature-Controlled Chambers

Many patch clamp experiments are performed at room temperature, i.e. between 20 and 24 °C, which is not only technically the easiest, but also allows particularly stable and low-noise recordings.

> ▶ **Caution:** In non-air-conditioned rooms, the temperature can change significantly between seasons! This can cause the "summer currents" to have a significantly faster kinetics and higher amplitude than the "winter currents", which of course disrupts any systematic series of experiments. However, constant air conditioning of the rooms also has its pitfalls, as the air flow from the air conditioners can affect the stability of the recordings. Measurements at elevated temperatures (30–32 °C) can therefore be advantageous if the room temperature rarely exceeds this value (even in summer), such that the target value can always be reached by actively heating the bath solution and the chamber.

Some experimental questions can only be answered close to body temperature, for example when you want to study the kinetics of membrane processes (postsynaptic currents, calcium influx, action potentials, etc.) under the most realistic conditions possible. In these cases, you need heating systems that can precisely control the temperature of the bath solution and the preparation. These are available from several manufacturers, but they are not cheap. Building your own with the help of a competent workshop is a good alternative.

The simplest method of temperature control is to heat the bath solution just before it enters the chamber and perfuse it at a steady, rapid rate. With a temperature sensor in the chamber, you can now adjust the in-line heater to maintain the desired temperature in the center of the chamber. It is important to ensure that either heat-resistant tubing (e.g. Teflon) or glass tubes are used at the heating surface to avoid damage. Heating is usually provided by either resistance wire (thermocouple wire) or Peltier elements. A close and well insulated contact between the heating element and the perfusion is important. More complex systems offer a combination of in-line heating and direct heating of the recording chamber, but this is only effective with metal chambers. The use of additional calibrated temperature sensors near the heater itself also allows for feedback temperature control that is less dependent on a constant flow rate. In practice, however, this does not guarantee a constant temperature during the experiment, but mainly serves to prevent the heater from burning out. Complex, heated chambers are not cheap, but homemade "water games" require some tinkering and can cause hum and switching problems.

4.2.3.3 Slice Chambers

It is often desirable to study cells in their native tissue association—especially in neurophysiology, but increasingly also in organotypically differentiated tissues derived from stem cells (organoids). In most cases, patch clamp studies on such preparations are performed in so-called submerged chambers, i.e. they are completely immersed in a large amount of extracellular solution compared to the tissue volume (Fig. 4.4a); the alternative model of an interface chamber is described in the next paragraph.

The study of complex three-dimensional preparations requires, in addition to the best possible optical conditions, stable storage of the specimen and above all an optimal supply of oxygen and glucose to the deep-lying cells. To fix the preparation on the bottom of the recording chamber, a small frame or ring made of platinum wire can be used, onto which nylon threads are glued (superglue). The ring or "harp" holds the sample in place but causes small incisions in the tissue. Alternatively, a brain slice or organoid can be glued directly to the bottom of the chamber or to a cover slip using poly-D-lysine. However, this requires some practice, since the glass bottom can only be kept dry for a short time while gluing the tissue, so that the specimen is not damaged.

As work on neuronal network oscillations has shown, native activity patterns are only obtained when the chamber is perfused at a very high flow rates (>6 ml/min) (Hájos et al. 2009; Maier et al. 2009). This suggests that the supply inside the preparation is critical and that the cells may easily lose their normal functions. Measurements of the rapidly decreasing oxygen partial pressure inside tissue sections have confirmed this. The situation is improved if the slice is not directly placed on the glass bottom of the chamber, but can also be perfused from below. This can be achieved by using nylon threads tightly stretched over a Plexiglass frame. (Nylon threads are most easily obtained from wax-free dental floss or women's stockings—we leave the choice of suitable models to the reader.) There are also commercially available chambers (double-perfusion or perforated chamber) with various designs for more efficient perfusion (Fig. 4.4b). Alternatively, the slices can be placed on filter paper or laboratory grade lens cleaning paper. This avoids cuts on the surface and protects the tissue. Caution**:** some papers are toxic to cells!

A special alternative is the interface chamber (Haas et al., 1979). Here, the preparation is placed on a permeable carrier (e.g., lens cleaning paper), at the

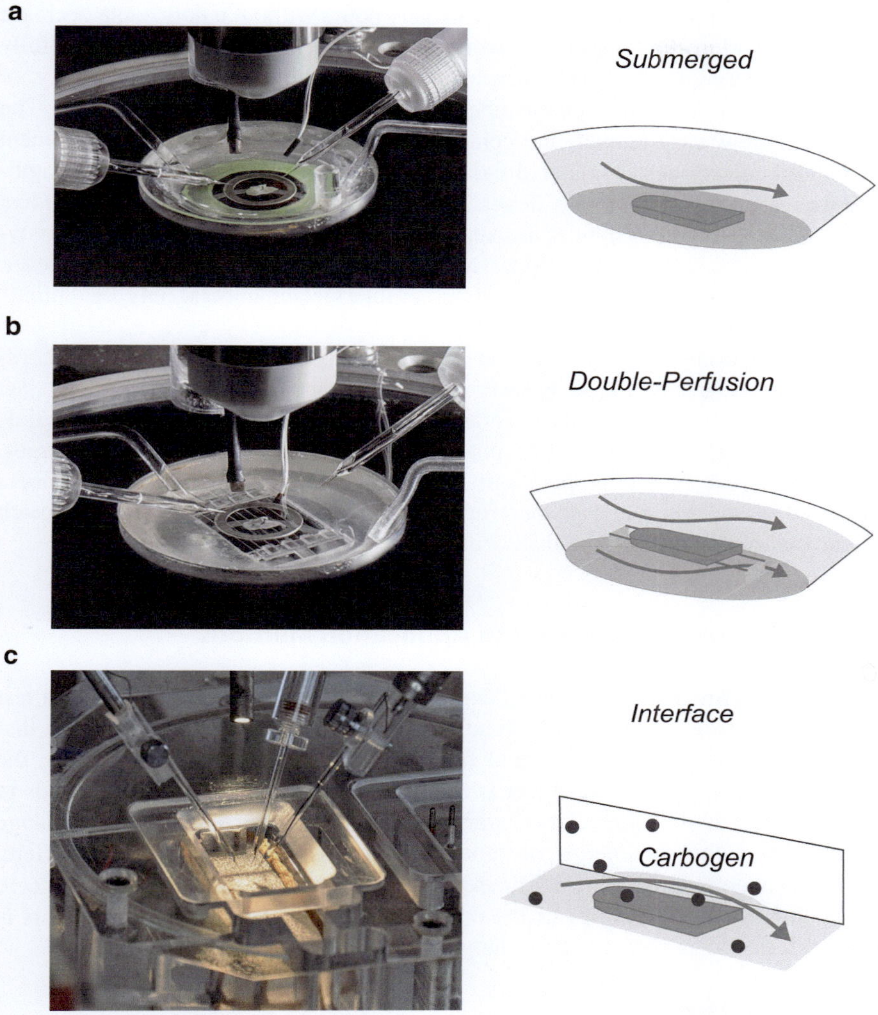

Fig. 4.4 Different types of recording chambers. **a** Standard submerged chamber; the specimen lies on a glass base. **b** Submerged chamber with a platform of nylon threads; the specimen is also surrounded by extracellular solution from the underside (double perfusion). **c** Interface chamber; the preparation is located in a moist environment saturated with carbogen and is only surrounded by a thin layer of extracellular solution

interface between the bath solution and the surrounding atmosphere. Therefore, it is only superfused by a thin film of liquid (Fig. 4.4c). A humidified and heated gas mixture, usually carbogen (95% O_2 / 5% CO_2) is supplied from above and from the sides. In our experience, this is by far the most tissue-friendly method for in-vitro measurements. These chambers are particularly useful in neurophysiology, allowing measurements to be made over many hours without a noticeable decrease in activity within the neuronal networks. It is often much easier to successfully characterize complex activities such as network oscillations, epileptic activity, or long-term potentiation in such chambers than in the submerged configuration. The chambers allow easy temperature control and offer a large opening for placement of multiple electrodes. A major disadvantage, however, is that the usable magnification is limited to that of a dissecting microscope. Visualization of individual cells or high-resolution optical measurements are therefore not possible (activity-dependent intrinsic optical or fluorescence signals at the network level, however, are possible). For patch clamp recordings in interface chambers, only the "blind" method remains.

Multi-electrode arrays (MEAs) can also be combined with patch clamp recordings. Such MEAs are offered by several companies and provide up to 4096 contact points in the tissue. This makes it possible to record extracellular field potentials or extracellularly recorded action potentials of individual cells (unit discharges, units) in complex tissue preparations (especially brain and heart). The chambers are open at the top and can therefore be combined with high-resolution patch clamp measurements of individual cells.

4.2.4 Chamber Perfusion and Application Methods

In many experiments, you want to expose the cells or membrane pieces to a test substance, a drug, or different bath solutions. When changing the solution, a distinction is made between slow and fast application. With slow application, the entire bath solution in the chamber is usually replaced over a period of seconds to minutes. Fast application refers to various methods in which the solution change takes place within an interval of about one second to a few milliseconds. Here, fluid exchange is usually locally limited to the immediate area of the patch or cell. In analogy to the fast jumps of the command voltage clamp experiments, this is often referred to as concentration clamp.

4.2.4.1 Slow Perfusion

The slow application is part of almost every patch clamp rig and is easy to set up. It is used to supply the cells with substrates, to modify membrane processes by specific drugs, or to determine the ion selectivity of channels. For this purpose, a flask (e.g., laboratory glass flask or an open 50-ml syringe) placed higher than the recording chamber is sufficient to supply the chamber quickly with solution through tubing by gravity. The flow rate can be easily regulated using hose clamps or by adjusting the height of the flask on a stand. However, so-called roller pumps are most commonly used, which ensure exact and constant perfusion rates regardless of the amount and position of the solution. Please note that this may introduce an additional source of hum or mechanical pulsations that must be compensated for by appropriate tubing material and careful grounding.

If you now want to change the perfusion solution, you either have to move the inlet of the pump to another reservoir, which often causes air bubbles and disturbances that can be fatal for sensitive recordings. Alternatively, you can combine standard clinical syringes (usually 10 to 50 ml) into a simple application system (Fig. 4.5) to quickly switch between several solutions with different substance or ion concentrations. As storage containers the syringes (without plunger) are hung next to each other on a stand or on the wall of the Faraday cage. A three-way valve is attached to the outlet of each syringe, releasing the desired solution and blocking the others. The individual tubings are either fed into the bath separately or

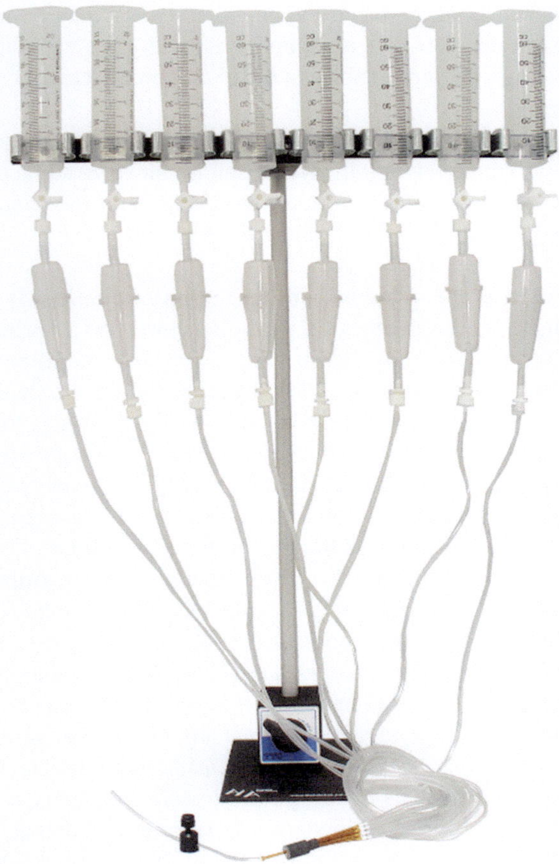

Fig. 4.5 Bath perfusion. In this simple version, eight different extracellular solutions are switched on or off manually (valves below the syringes). The drip chambers decouple the solution supply from the recording chamber and can reduce the line noise (hum) (image courtesy of npi electronic, Tamm, Germany)

combined in a common inlet, but this can easily result in some mixing of the solutions. Therefore, unneeded solutions should always be shut off with a valve.

On the other side of the bath, the solution must of course be drained, usually by means of a cannula or a glass tube to which a vacuum is applied. This is often somewhat delicate and can cause problems. If the suction fails for a short time, the recording chamber, stage, and microscope are quickly flooded with the salt- and sugar-containing solution, which then requires extensive cleaning (some manufacturers offer protective films for the condenser and microscope stand, but flooding should still be avoided). In addition, if the aspiration is poorly adjusted, the contact between the drain and the extracellular solution can be repeatedly broken. When the aspiration loses contact with the solution, it causes vibrations and fluctuations of the water level in the recording chamber, which may destabilize the seal and cause electrical artifacts. We recommend curved glass capillaries for aspiration (Fig. 4.4a). Others prefer metal cannulas or diagonally cut tubes.

▶ For good drainage, it is helpful to coat both the chamber and the transition to the aspirator with a thin layer of salt to minimize surface tension. You can do this by letting a little salt solution dry on the chamber and aspirator overnight before using them the first time.

Depending on whether the solution is to be reused (recirculated) or always discarded, the vacuum can be generated by a roller pump or a vacuum pump (diaphragm pump or compressed air). If you want to use the same roller pump for both inflow and outflow, the pump tubing for draining should be significantly larger than the one for inflow. Alternatively, you can use T-hose-fittings to use two pump hoses in the draining line to double the suction capacity. Diaphragm pumps (such as those used in aquarium air pumps) are not suitable for contact with the solution and must be used in combination with a vacuum bottle. These can be purchased ready-made or you can drill two small holes in the plastic lid of a laboratory glass bottle and glue in two suction nozzles (from plastic pipettes). **Important:** Empty the bottle regularly and check the vacuum at the recording chamber afterwards!

In interface chambers, the solution is usually drained by gravity through a large hole at the end of the recording chamber (a fine strip of filter paper or nylon placed between the edge of the chamber and the drain nozzle prevents meniscus formation). The liquid is then simply drained through a hose into a waste container.

It is recommended to measure the bath exchange times at the beginning of a series of experiments. When has my drug really reached the cells at full concentration? Depending on the flow conditions in the recording chamber, you will need to exchange five to ten times the volume of the chamber, even under simple conditions. You can roughly estimate the time required in a preliminary experiment without cells by using colored solutions (food coloring or trypan blue). However, it is more accurate to measure the change of liquid junction potentials (Sect. 3.2.3) with a patch pipette immersed in the bath or even to perform a test recording in the desired patch clamp configuration using a solution with a clearly measurable effect (e.g., changing the potassium concentration when measuring the membrane potential). In slice preparations, one should not underestimate the slow convection of the solution into the interior of the slices and measure it as described above! In interface chambers, the exchange is even slower than in the submerged situation. Here, the equilibration times of low-molecular substances inside 400 μm thick brain slices can be up to 1 h!

In many pharmacological experiments, one wants to wash in substances and then wash them out again to check whether an observed effect is reversible. This is an important control to see if the cells were not damaged during the recording, and the apparent drug effect is just a non-specific artefact. Typically, the "wash-out" of a substance takes significantly longer than the "wash in", because many substances accumulate in the tissue, especially if they are lipophilic. In addition, the affinity to the target structures is important, so that e.g. high-affinity toxins often cannot be washed out at all or only very poorly. Often, the reversal of a pharmacological effect is not fully achieved during the stable "lifetime" of the patch clamp recording, and in most series of experiments it will not be successful in all recorded cells.

▶ **Tip:** Always make sure that there are no residual substances in the tubes, chamber or tissue before starting the next measurement!

4.2.4.2 Fast Perfusion

The analysis of very fast substance effects, for example the activation of ligand-gated ion channels, requires a correspondingly fast perfusion system. The solution change at the membrane should be significantly faster than the time constant of the fastest change to be investigated. For this purpose, several more or less locally acting application methods have been developed, which work either mechanically (through flow), electrophoretically or by optical activation of substances. For whole-cell recordings, especially with branched or sensitive cells, mechanical applications can achieve minimum times of a few tens of milliseconds, while for very compact cells or isolated patches, solution changes in the range of one millisecond or less are possible. Many systems are commercially available, but one can also self-build useful fast application systems. Figure 4.6 shows a widely used principle.

Mechanical methods work with flows that are as laminar as possible and are directed at the cell or the patch. For example, you can buy (or build) a perfusion system, in which many supply lines end in a common, small tip that is positioned close to the cell. By quickly opening and closing valves, the solutions are now switched between. Such arrangements are used in many variations for whole-cell recordings (mostly on cultured cells). Even faster is the application of substances to isolated patches, which is achieved by piezo-driven lateral movement of two laminar flows (Fig. 4.6). For this purpose, theta glass with two barrels separated by a thin glass layer (septum) are used allowing application times well below one millisecond (Sylantyev and Rusakov 2013; Danker et al. 2016).

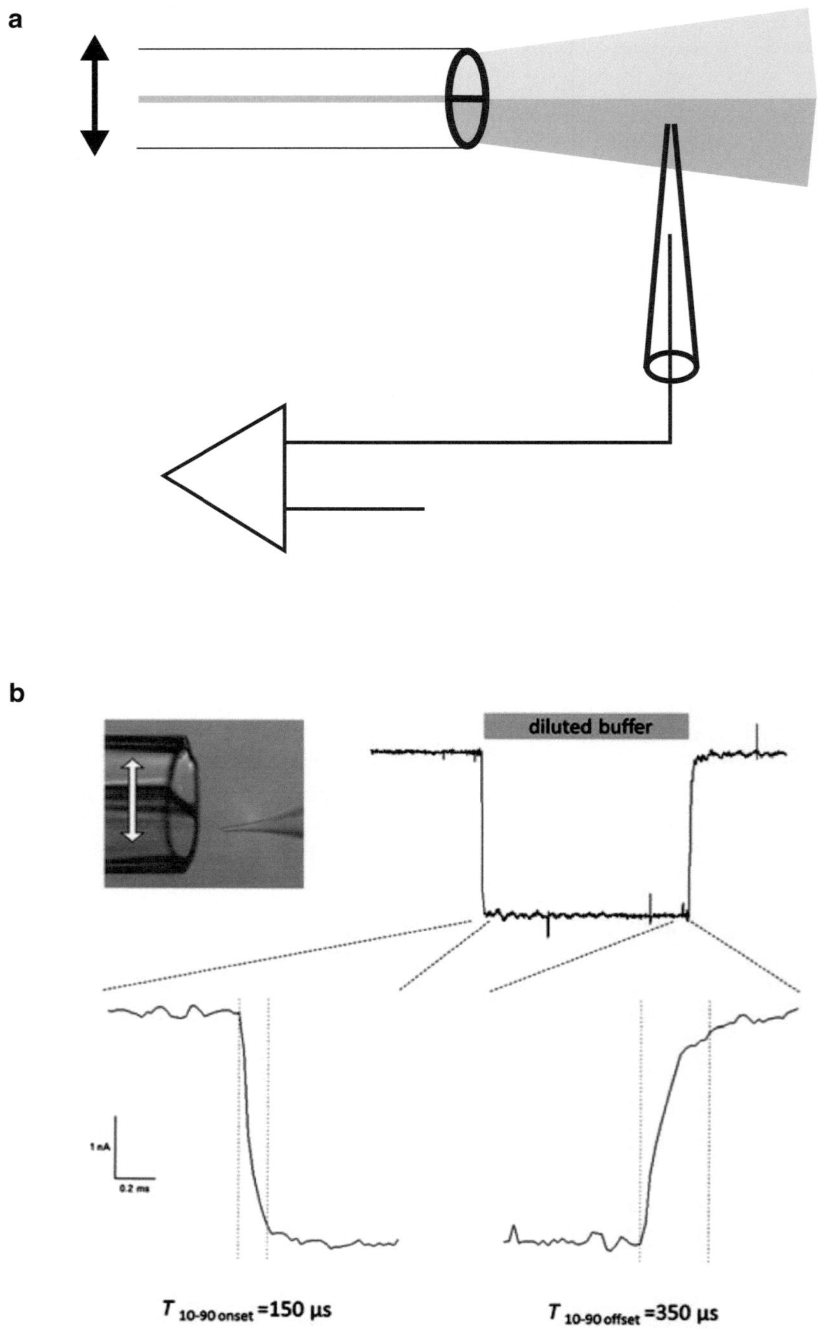

Fig. 4.6 Rapid substance application. **a** Principle of application using a theta glass. The pipette tip protrudes into the laminar flow and can be exposed to one of the two solutions by rapid movements of the glass. **b** Measurement of the application speed (here with the pipette in a different spatial position). Application speed was about 150 µs (on) and 350 µs (off) in this recording. (Modified after Danker et al. 2016)

Charged substances such as glutamate can also be ejected from a pipette using DC voltage potentials. This method is called iontophoresis or microiontophoresis and allows for a very fast local application below one millisecond with a suitable pipette and high-quality control device. For details (and for the selection of suitable substances) we refer to the literature.

Another modern method to achieve fast substance application is the photolytic release of caged compounds („uncaging“). The substance to be applied is provided with an encapsulating compound making it biologically inactive. The cage can be removed by a strong flash of light (in the UV or also in the visible range), so that the substance is activated very quickly. Photolytically activatable transmitters, as well as intracellular second messengers such as cAMP or IP_3 are commercially available. They can be added to both the bath or to the pipette solution. For activation, a flash lamp with a high UV component can be used, whose beam can be focused through the optics of the microscope. Alternatively, a laser may be used. A particularly focused application is possible with two-photon microscopy, which allows very fast and localized release of activated molecules, e.g. near single synaptic spines (Kantevari et al. 2010).

4.3 Electronic Components

4.3.1 Pre-Amplifier and Amplifier

The patch clamp amplifier usually consists of a pre-amplifier and a main amplifier or control unit. The pre-amplifier (called headstage or probe) is the heart of the electronics, where the electrical signals from the cell are measured and currents are injected. These are read out as a proportional voltage signal; the device is therefore referred to as current–voltage converter or I-V converter. Currents and voltages in patch clamp experiments are in the order of pA (10^{-12} A) to nA (10^{-9} A) and from 1 mV (10^{-3} V) to about 100 mV. These small signals are easily be lost in noise, cannot be transmitted over long distances without stabilization, and should not be distorted by the measurement itself. The highly sensitive pre-amplifier, which first picks up the tiny signals, is therefore placed as close as possible to the recording electrode. There, the signal is amplified and stabilized before it travels a longer distance to the main amplifier.

The pre-amplifier must be mounted stably and free of vibrations. For example, the housing of the headstage can be screwed directly onto a plate on the micromanipulator. This plate should be made of a light, but strong and non-conductive material (plastic), as the metal parts of the pre-amplifier should not be in contact with the metal parts of the manipulator. The cable leading from the pre-amplifier to the main amplifier must definitely be fixed to a stable structure (column or cage) near the pre-amplifier (strain relief).

The main amplifier or control unit is usually located in a rack next to the anti-vibration table. This is where the signals are further amplified, filtered, and digitized. This is done either in the amplifier itself or in a separate analog-to-digital converter (ADC), to which the analog signal from the amplifier can be coupled.

All control signals are either controlled by software via the computer or set via front panel switches and potentiometers on the main amplifier. There, they are converted into analog signals for the pre-amplifier. Usually, a special control program is used. It allows the display of the measured voltages or currents, the setting of the amplification factor, the switching between different measurement configurations, and often also the storage and a first analysis of the data. Most of the work is therefore done from the computer, while the use of selector switches and dials on the amplifier itself are becoming increasingly rare. For the older generation, the haptics of a "real" device is reassuring because it conveys a sense of understanding and control. For younger readers, this idea is probably as distant as traveling in a horse-drawn carriage. Nevertheless, an electrophysiological experiment is not a computer game, and it is important to always know how the various settings in the program specifically affect the measurement. We strongly recommend going through all configurations of the control program using a model cell until you really feel confident.

Due to the extremely high sensitivity of the pre-amplifiers used for the patch clamp technique, it is particularly important to ground all conductive surfaces near the headstage's input. For that, the pre-amplifiers have a ground input, which is often connected to the reference potential of the amplifier. This can be used to connect nearby parts of the setup, such as the microscope and the bath electrode. Alternatively, only the bath with the specimen is grounded via this input and all other conductive parts of the setup are grounded through a separate ground line.

Friction on clothing or furniture can cause very high electrostatic charges on the skin (several thousand volts). This is especially true in dry air, e.g. in winter. If you touch the input of the pre-amplifier with this charge to change the pipette, it can destroy the high-impedance input of the pre-amplifier. In the worst case, all subsequent measurements will be corrupted without you noticing. At the very least it is an annoying, time-consuming and expensive affair. Therefore, you should ground yourself during the experiment or at least before each pipette change. Special high-impedance grounding wristbands sold at electronics stores are the most convenient way to do this.

▶ **Caution:** It is a fatal mistake to simply touch the grounded tabletop of the rig or a ground wire for grounding yourself, especially when you are tinkering with any component! This creates a low-resistance connection to the ground conductor, which can cause sudden death if you accidentally touch an AC power source with your other hand! For safety reasons, this is prohibited anyway, but we want to make you aware of it again.

4.3.1.1 Choosing an Amplifier

There are several commercially available patch clamp amplifiers. They operate according to one of three possible principles: via a feedback resistor R_f, via a capacitive feedback C_f, or with the discontinuous SEVC (Single-Electrode Voltage Clamp) method.

We have explained these principles in more detail in Sect. 3.1. Compared to traditional amplifiers with feedback resistance, capacitor feedback amplifiers have advantages due to their lower noise, especially for high-resolution recordings (single channels), otherwise both types are equally suitable. SEVC amplifiers are particularly well suited for the use with high-impedance pipettes or even sharp microelectrodes. They can inject large currents and have good current clamp or bridge-mode characteristics at the same time. However, they do not allow single-channel recordings. Some amplifiers can also switch between different methods, such as resistance feedback (for whole-cell recordings or large single-channel currents) to capacitive feedback (for high-resolution single-channel measurements). Hybrid devices combining SEVC and high-resolution feedback methods are also available.

All commercially available amplifiers have the necessary components for most patch clamp experiments. However, there are a few points to consider when buying a new amplifier:

- How large are the signals that I want to measure? High-resolution measurements (e.g. on single channels with low conductance) require a low-noise amplifier, while large currents (e.g. voltage-activated sodium currents in whole-cell mode) may require an amplifier with particularly large output signals.
- Do I want to use the same amplifier for patch clamp measurements as well as other techniques? These include, for example, extracellular field potentials or conventional intracellular recordings.
- It is also important whether you want to record "only" with one electrode in the long term or, for example, want to perform paired recordings from two or more cells. In this case, you may want to purchase an appropriately equipped amplifier with multiple pre-amplifiers.

All these conditions should be carefully considered. Then you should make your choice in discussion with colleagues and suppliers.

4.3.2 Intermediate Amplifiers and Filters

Modern patch clamp amplifiers usually have built-in filters to reduce the high-frequency noise in the signals. These are called low-pass filters because they allow only frequencies below a cutoff frequency to be recorded. Additional, intermediate filters built into the signal path allow sharper and more variable processing, but are often unnecessary (see below). You should always be aware of how the respective filter changes the original signal—to be on the safe side, you should record the unfiltered signal in parallel. For example, if the recorded signal contains very fast components (action potentials, fast activating voltage-dependent or synaptic currents, single-channel currents), you must not filter too low, because this would weaken and slow down the steep flanks.

On the other side of the spectrum, you may in some cases want to prevent a slow drift from gradually pushing the measured signals out of the measurement range. To do this, one uses a high-pass filter that "cuts off" fluctuations below a certain frequency. This manipulation is not harmless either—we recommend trying it out once with different cutoff frequencies in real measurements. If the biologically interesting signal contains slow components (e.g. a longer plateau phase of a current), their shape can be significantly distorted by a high-pass filter!

To be absolutely sure, it often makes sense to filter rather conservatively during data acquisition and use appropriate digital filters post hoc in a suitable analysis program. These changes to the data are—in contrast to the filtering of the original measurement—reversible and therefore harmless. If you combine your amplifier with an external A/D converter, you can also split the analog output and record the signal twice with different filters. Again, the less strongly filtered data set is an important backup that contains the maximum information about the raw data.

There are very good external filter devices that can often also additionally amplify the signal. However, it is best to set the amplification factor for the measured currents on the patch clamp amplifier as high as possible, as this leads to a better signal-to-noise ratio. A frequently used external intermediate filter is the "HumBug" or similar main noise eliminators. These are filters that selectively remove the 50/60 Hz hum of electromagnetic induction. Here, the noise suppression is actively synchronized to the mains frequency, preferably without affecting measurement signals in the same frequency range. We are rather reluctant to use additional filters and recommend that the noise/hum in the unfiltered signal should be eliminated as much as possible (Sect. 5.3). If you do use a HumBug, you should definitely perform a control experiment and evaluate relevant signals once with and once without the filter.

The advice given for the HumBug also applies in general: To avoid unpleasant surprises after completing a series of recordings, it is advisable to measure the signals of interest several times with different filter settings at the beginning and to evaluate the results once as a trial. You will then have a good basis for deciding at which frequency you want to filter. The same applies to the appropriate sampling frequency (also called sampling rate), which we will discuss in Chap. 5.

4.3.3 The Stimulation Device

In neurophysiology, it is often necessary to electrically stimulate cells or axons from the outside. For this purpose, either a glass pipette or a metal electrode (monopolar stimulation), or two closely spaced platinum or tungsten electrodes (bipolar stimulation) are used. In order to prevent hum from entering the rig via the connecting cables, stimulators are used that are "galvanically isolated", i.e., their output stage has no direct contact with the power supply, through which the external alternating voltage is fed. Such isolation is achieved with coils, capacitors or optical components and reduces the risk of picking up massive mains hum via the stimulation cables. Alternatively, battery-powered devices are also recommended. Stimulators can be externally triggered by the control program from the outside and should provide short rectangular pulses of less than one millisecond duration (ideally 100 or even 50 µs) and up to several tens of volts (50–100 V). Suitable stimulators are available from many suppliers of electrophysiological equipment.

4.4 Glasses, Pipettes, Electrodes, and Solutions

In this section, we want to address the whole process of making pipettes: from the different types of glass to the necessary equipment to make the pipettes, to filling the pipettes with electrolyte solution and treating the electrode wires.

4.4.1 The Patch Pipette

The tip of a patch pipette is characterized by its flank shape (short and blunt or long and pointed) and the size of the opening. The art of pipette making is usually to precisely control these two properties independently of each other. Both ultimately determine the capacitance and the electrical resistance of the pipette. This depends not only on the tip opening, but also on the shape of the flanks. The longer the pipette tip is the greater the "flank resistance" (Fig. 4.7).

Thick-walled pipettes (wall thickness of the capillaries about 0.3–0.5 mm) are clearly better suited for single-channel recordings due to their lower capacitance, as this reduces the noise in the patch clamp recording. For this reason, the pipette's wall is sometimes artificially thickened with coating (see below). Also, the diameter of the pipette opening is often optimized for certain measurements—e.g. for single-channel recordings of a rare channel, larger pipettes are used to increase the "hit rate", while small pipettes can be used for recordings of a channel present in high density. In the latter case, having multiple channels open at the same time may even disturb the evaluation of the data.

Thin-walled pipettes (wall thickness of the capillaries about 0.15 mm) are mostly used for whole-cell or perforated-patch recordings (Sect. 6.2), because here it is important to have as low a resistance as possible. They are also easier to produce in the desired shape. Pipettes with an opening diameter between 0.5 and 1 µm have low resistances between 2 and 5 MΩ. In practice, however, one will rarely try to determine the pipette diameter, but rather optimize the pipettes simply based on the measurable resistance.

▶ The lower the resistance of the pipette, the lower is usually the series resistance. However, with small cells, it is difficult to get a good seal with large, low-resistance pipettes—you have to test different shapes until you find an optimal compromise.

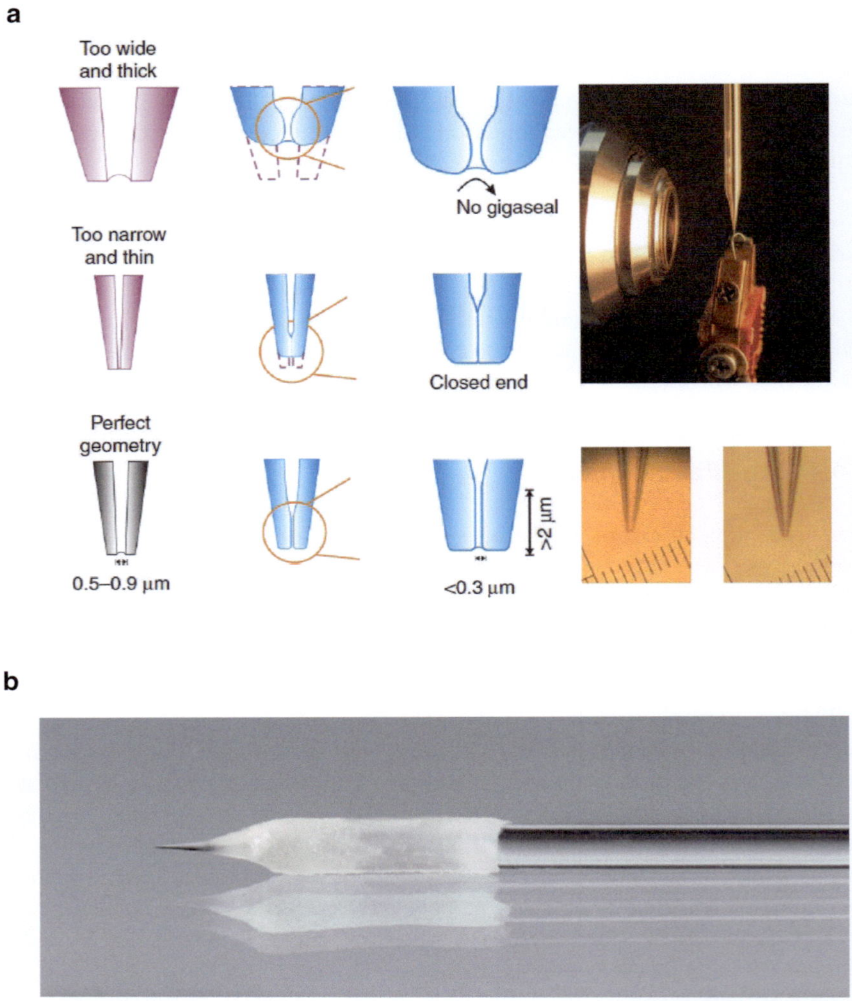

Fig. 4.7 Pipette shapes. **a** Different shapes of the pipette tip before (left) and after (center, right) fire polishing. Photos on the right: Objective and glow wire of a microforge (Fig. 4.8) with pipette nearby. **b** Pipette coated with wax. (**a** from Chen et al. 2017)

While the suitable resistance value of the pipette depends on the application, all patch pipettes should have the lowest possible capacitance to minimize the "background noise". This property depends on both the wall thickness of the pipettes and the glass used. However, capacitance and background noise can also be improved by additional measures such as coating with hydrophobic substances (see below).

Most glass capillaries used in the patch clamp technique have an outer diameter of 1.2–2 mm and an inner diameter of 0.5–1.5 mm, depending on whether thin-walled or thick-walled pipettes are required. An outer dimension of 1.5 mm is widely considered the standard and offers a good compromise between mechanical stability (important for long pipettes that must protrude far from the holder) and electrical properties. Thicker capillaries (2 mm outer diameter) are even more stable, especially in combination with a high-quality pipette holder with two clamping points. They also provide more material for shaping the tip, making it easier to produce pipettes with large tip openings. However, the larger surface area increases the capacitance, which increases the noise level. These factors must be considered together when selecting the glass capillaries and the appropriate pipette holder, as high-quality and mechanically stable pipette holders are only built for a certain diameter. Similarly, not all pipette pullers (see below) allow the use of thick capillaries.

4.4.1.1 Pipette Glasses

Patch pipettes should be made fresh from glass capillaries and not too long before the experiment. The glass capillaries can be purchased ready-made from laboratory suppliers or directly from glass producers, who may also offer special sizes. Capillaries with fire-polished ends are of great advantage for patch clamp applications, as they protect the electrode wire and rubber seals when changing pipettes. The sealing properties of different types of glasses not only vary from glass type to another, but can also vary between different manufacturers of the same glass type. In case of problems, it is recommended to test capillaries from different manufacturers. However, this is laborious, so all other possible errors should be ruled out first, which we will discuss in Sect. 5.1.3. The glass from which the patch pipettes are fabricated must meet four requirements:

1. It should be easy to pull into pipettes.
2. It should easily form seals.
3. It should generate as little electrical noise as possible.
4. It must not be cytotoxic.

For the first two points, the melting point of the glass is most important; for the noise properties, the dissipation factor and the dielectric constant are decisive. The chemical composition of the glass needs to be considered for cell toxicity.

Types of Glass
Glass is distinguished according to the melting point:

- Soft glasses (soda glasses, below 700 °C)
- Medium-hard glasses (borosilicate, 700 to 850 °C)
- Hard glasses (aluminum silicate, above 900 °C)

For particularly low-noise recordings, the very hard quartz glass (melting point at about 1600 °C) is also used, which requires a special pipette puller.

Originally, soft glasses were used for the patch clamp technique. They are easy to pull due to their low melting point and result in short, thick pipettes with steep flanks and a large tip diameter, yielding low pipette resistance. They are also easy to heat polish and have excellent sealing properties. However, due to their sometimes poor electrical properties (high noise), they are no longer used as patch clamp pipettes. Nowadays, medium-hard **borosilicate glasses** are mainly used for patch clamping. In practice, they offer the best compromise between precise adjustment of the pipette shape and good electrical properties. The hard **aluminosilicate glasses** are rarely used but have the best noise properties of all the glasses mentioned so far. However, they are also the most difficult to pull and generally form poorer seals.

Of all the glasses used to make pipettes, **quartz glass** has the best electrical properties. In addition to the extraordinarily low noise, quartz pipettes also have another advantage: quartz glass does not contain any additives that could lead to contamination. However, quartz glass has two disadvantages: it is expensive, and it has a very high melting point, which is why pipettes from quartz glass cannot be pulled with conventional pullers.

Special devices have been developed for pulling quartz pipettes, in which the glass is not heated by a heating filament, but by a CO_2 laser. However, these laser devices cost about twice as much as conventional pullers. Therefore, quartz pipettes are only used for extremely low-noise single-channel recordings, where, of course, the noise from all other sources should also be kept as low as possible. Thick-walled quartz glass capillaries (outer diameter 1.5 mm, inner diameter 0.6–0.75 mm) are then used, and the quartz pipettes are additionally coated (see below). This high (also financial) effort will only be necessary in special situations.

4.4.2 Pipette Production

The glass capillaries should be free of dust and grease. In the past, they were laboriously cleaned by ultrasonic treatment in alcohol or laboratory detergent and then dried out at 200 °C for about 30 min to remove residual moisture (this also reduces the noise caused by humidity.) Today, this is only done in exceptional cases, for example, when dust or moisture problems occur. Such contaminations of the pipette tip are clearly visible under the microscope during the recording. However, the possibility of cleaning and drying the glass should be considered in case of persistent sealing problems (Sect. 5.1.3).

4.4.2.1 The Pipette Puller

A pipette puller is used to fabricate the pipettes from suitable glass capillaries. It has evolved from relatively primitive predecessors to a high-tech device that allows multi-stage program sequences for pulling blunt pipettes with almost any opening diameter. The art here lies in the precise control of the pulling force, the heating, and the exact position and extent of the heat.

Depending on the direction in which the capillaries are mounted, there are vertical and horizontal pullers. With both types, at least the power supply of the heating element (and thus the heat) and the duration of the individual pulling steps can be adjusted. With conventional vertical pullers, very usable pipettes can be produced with a maximum of two pulling steps with some practice. However, the more expensive programmable horizontal pullers are now most common.

One of the advantages of a programmable puller is that several people can conveniently use it to make different pipettes. For example, someone recording in a slice preparation needs much slimmer and longer tips due to the short working distance between the objective and the specimen than someone working with cell cultures using an inverted microscope, with plenty of space from above. There are also applications that require particularly high or low resistance pipettes. In these cases, pullers with multiple stored programs are very useful.

It is important to handle the filament very gently, as even the slightest touch can bend the heating filament, rendering it useless for well-defined pipette shapes. Colleagues are usually not happy about this. For a similar reason (consistency of parameters), pipette pullers should not be placed near windows, heaters, air conditioners, or any air currents.

4.4.2.2 Pulling the Pipettes

A patch pipette is usually pulled in at least two steps, with horizontal pullers usually in four to five steps. In the first step, the capillary is "pre-pulled", i.e. it is heated and extended by about 0.7–1 cm. This makes it about 0.2–0.4 mm thin in the middle. The cooled pipette is then re-centered in vertical pullers, its thinnest point is brought into the center of the heating coil, and then fully extended. This causes the capillary to split at its thinnest point, producing two pipettes. With horizontal pullers, on the other hand, the re-centering may be skipped, and the pipette is simply pulled out further until separation in one of the following steps.

The first three fabrication steps can be important for the length of the tip, e.g. for making long thin pipettes for use under immersion objectives. The parameters of the last pulling step, such as heat, pulling force, and pulling speed, must be set very precisely, as these settings have a significant influence on the shape and properties of the pipette opening. The choice of appropriate parameters for pulling pipettes depends on many factors and has to be worked out on a case-by-case basis. However, as a rule of thumb: the greater the heat and the pulling speed, the smaller the diameter and the flatter the flanks of the pipette.

4.4.2.3 Polishing with the Microforge

Originally, the tip of each patch pipette was polished in a microforge immediately before use to smoothen and thicken the glass at the tip. Nowadays, most users skip this step and still achieve very good recordings with the slow and gentle processing of modern pullers. However, for some pipette shapes and preparations, or

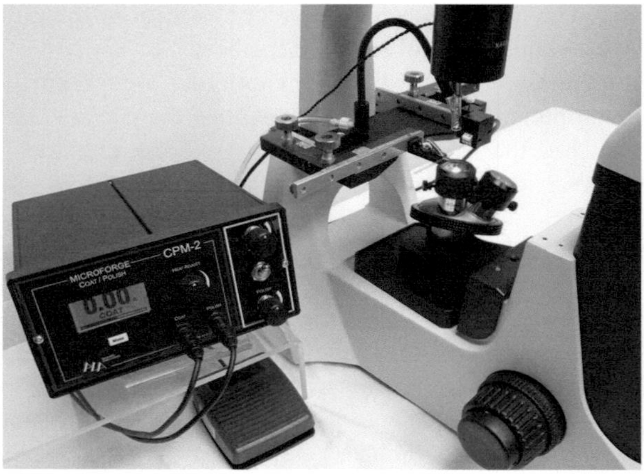

Fig. 4.8 Microforge. The core components of the system are the microscope (right) with pipette holder and the control unit for the current that determines the heat of the annealing filament near the pipette tip. In this model, there is also a hot air source for drying the pipette coating. (Image courtesy of ALA Scientific Instruments, USA/npi electronic, Tamm, Germany)

when maximum seal resistance is required, the pipette should still be polished. Polishing may also be necessary after certain coating techniques (see below).

The microforge (Fig. 4.8) is a microscope with a pipette holder and micromanipulator, in which the pipette tip is brought under visual control very close to a heating wire, which is heated by a controllable transformer. Usually, a small glass ball is melted on and cooled by an air stream, creating a steep temperature gradient, which is used to melt the tip of the pipette. These devices are available from specialized suppliers.

4.4.2.4 Coating the Pipettes

Pipettes can be coated with hydrophobic substances to increase the wall thickness of the pipette tip and reduce wetting with liquid. This reduces the electrical capacitance and therefore the noise. Coating is particularly important for pipettes that are brought into the bath at a flat angle. The most commonly used coatings are silicone elastomers (e.g. *Sylgard 184*, Dow Corning Corp.), various waxes (beeswax, dental wax) or other plastics. In the best case, a coating with Sylgard halves the background noise of borosilicate glasses, which can then reach the quality of aluminosilicate glasses. In general: For single-channel recordings, the pipettes should be coated with a hydrophobic substance as close to the tip as possible; for whole-cell recordings, it is often sufficient to prevent a water film from being pulled up on the pipette.

Coating with Sylgard requires treatment in the microforge as described above. To prevent the Sylgard from running to the tip, the pipette is fixed with the tip pointing slightly upward. For coating, you can use, for example, a small glass hook or a bent cannula. It is very important that the very tip of the pipette (about

10–50 μm) remains free of coating, otherwise no seal will be formed. The Sylgard is then hardened by the heating coil or by a stream of hot air.

> **Tip** A freshly prepared Sylgard mixture (mix well!) can be stored in a sealed container in the freezer for several weeks. Once thawed and opened, it will polymerize within a few hours. Before opening, the Sylgard must be brought to room temperature to prevent condensation.

Coating with wax is somewhat easier: hold the pipette tip upwards and apply the wax, heated just above the melting point, with a spatula while rotating the pipette. The wax is applied to the flank a few millimeters below the tip and then spread down as far as the pipette will later be immersed into the bath solution (Fig. 4.7b).

Quartz pipettes should always be coated to take full advantage of the low noise of this material. After coating with Sylgard, quartz pipettes must also be "heat polished". Although this does not melt the quartz glass in a conventional microforge, it does remove any remaining Sylgard.

4.4.3 Filling the Pipette

The finished patch pipettes should be stored in closed dust-free containers and used within a few days. Before the experiment, they must be filled with electrolyte solution (pipette solution, intracellular solution; Sect. 4.4.5). The solution should be filtered through a sterile filter (commercially available syringe filters) before filling. The solution can be stored either in a 1-ml syringe behind the filter or in a clean reaction tube, cool and protected from light.

Pipettes are usually filled from the back (backfilling) by means of long and thin pipette tips (microloader), available from laboratory suppliers, which can be connected directly to 1-ml syringes. Instead of a syringe, a 20–200-μl automatic pipette can be used.

> **Warning:** Do not use metal needles for filling, as this may contaminate the pipette solution with metal ions!

At least in long drawn-out patch pipettes, air bubbles often form, which are hardly visible and difficult to remove from the tip. Sometimes, they are only noticed when the resistance of the pipette is extremely high for no apparent reason. They can usually be removed by patient and gentle tapping. It is better to use glass capillaries with a built-in filament, so that the solution will automatically flow into the tip due to capillary forces. The pipette can also be filled by placing the tip in a tube with pipette solution and applying a vacuum to the back of the pipette (e.g. via tubing connected to a 10-ml syringe). For this purpose, the pipette can be mounted on a small stand and the tip fills bubble-free from the front (tip filling). The remaining solution is then filled into the pipette from the rear as described above.

The pipette should only be filled as far as the electrode wire mounted to the pipette holder dips into the liquid by a few millimeters. Overfilling the pipettes can have a negative effect on the noise properties, as there is a risk of the liquid being pushed into the pipette holder and forming a conductive bridge. For the same reason, droplets of electrolyte solution adhering to the pipette shaft are to be removed with a syringe. Despite all these precautions, the pipette holder should occasionally be disassembled, thoroughly cleaned and above all, dried (e.g. in a vacuum desiccator).

4.4.4 Electrode Wires

The pre-amplifier is connected to the pipette solution by a metallic conductor and to the bath solution by the ground wire. To avoid offset potentials (Sects. 3.2.3 and 5.1.1) or at least to keep them constant, only chloridized silver wires should be used.

4.4.4.1 Chloriding of a Silver Wire

The pipette holders are usually supplied with cleaned and cut silver wires, which are often already mounted. If you use a separately purchased silver wire, it should be sanded with fine sandpaper, cleaned with alcohol and dried. The wire is then immersed in a potassium chloride solution (about 1–3 M) and connected to the anode (positive pole) of a DC source. Another metal wire is connected to the cathode (negative pole) and also immersed in the bath (Fig. 4.9a). Chloride anions are “drawn” to the silver by the positively charged anode, which is why silver chloride is deposited on the wire. About 1 mA is used for chloriding; using a lower current for a longer time leads to better and more durable coatings.

Alternatively, the silver wire can also be chloridized by immersing it in concentrated chlorine bleach (sodium hypochlorite, NaClO) overnight—a simple procedure that gives good results. However, the chloride layer remains relatively thin.

▶ **Tip:** You can draw the bleach into a 1-ml syringe and carefully dip the wire into it. This can be done even if the wire is already mounted in the pipette holder.

A uniformly chloridized electrode has a matte, dark gray coating. Since the silver chloride layer of the electrode wires is often scratched by changing the pipettes (preferably use capillaries with polished ends!), the wire has to be changed or re-chloridized from time to time, at the latest when a significant drift in the pipette potential is observed. However, the drift can also be caused by the bath electrode. **Important:** Always store chloridized silver wires protected from light!

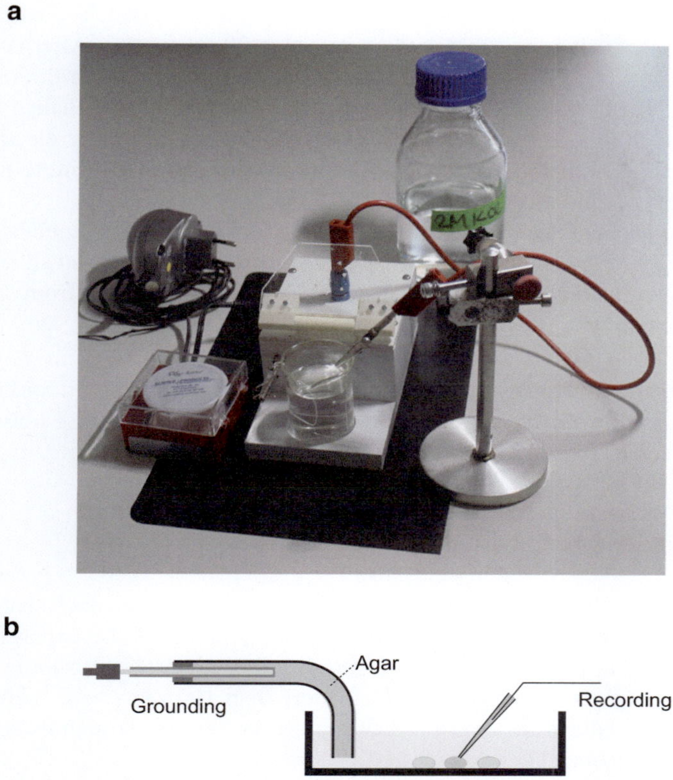

Fig. 4.9 Chloriding of silver wires and grounding. **a** System for chloriding silver wires. A voltage is applied to the wire using the voltage source (grey box) for 5–10 min according to an optimized protocol, whereby chloride ions accumulate at a positive potential. **b** Agar bridge for grounding the bath in experiments with changing ion concentrations

The bath electrode

A silver/silver chloride electrode also serves as a bath electrode. The potential difference between the bath and the recording electrodes should be zero, when both are immersed in solutions of the same chloride concentration. However, if the chloride concentration in the bath changes, a potential difference between the electrodes will occur, because the wire in the pipette is usually in contact with a different pipette solution. That's why many experimenters connect the bath electrode via a so-called agar bridge to the bath solution. (Fig. 4.9b). This allows the chloride concentration to be kept constant in the immediate vicinity of the electrode. It also avoids the direct immersion of the wire, as silver ions are toxic to many cells.

To make agar bridges, fill a perfusion tube or the tip of a cut-off Pasteur pipette with agar. Use about 3% agar–agar (note: percentage may vary by batch) in the bath or salt solution (e.g., 150 mM NaCl). The agar is heated, filled into the tube or pipette, and after cooling, topped up with the same solution. You can also fill a larger tube as much as possible with agar using a 20-ml syringe. After cooling, the tube is cut into pieces, which can be stored in the salt solution in the refrigerator for several weeks. Avoid KCl solution as it may unintentionally change the potassium concentration in the bath! Then insert the chlorinated silver electrode from the back and seal the tube with wax. Finally, connect the silver wire to the ground input of the pre-amplifier via a cable.

You can also use the somewhat more stable and durable AgCl pellets as a bath electrode. These are small pieces of a sintered mixture of silver and silver chloride (1:3) at the tip of a chlorinated silver wire that can be purchased from specialized laboratory suppliers.

▶ **Caution:** Incorrectly manufactured bath electrodes or agar bridges that are too thin can have considerable resistances and then cause noise or even an increased series resistance.

4.4.5 Intra- and Extracellular Solutions

Both the extracellular and intracellular environments must be such that cells and tissues can maintain their homeostasis for the duration of the experiment. This includes osmotic pressure, suitable composition of electrolytes, supply of essential nutrients (glucose, oxygen) and pH. More than 100 years ago, "cell-friendly" extracellular media were developed that preserve cellular properties such as the contraction of isolated hearts.

In patch clamp recordings, the inside of the pipette comes directly into contact with the membrane or the inside of the cell. Therefore, the pipette solution must be composed to maintain cellular integrity. Note that the pipette must simulate the *intra*cellular environment for whole-cell and for outside-out measurements, but the *extra*cellular environment for cell-attached or for inside-out measurements.

Of course, standard recipes have been developed over the years, which have been tested and optimized in thousands of experiments. However, they vary considerably between different preparations and target parameters of the experiments. It makes a big difference whether you want to measure in cultured muscle cells or in brain slices, whether you study excitatory or inhibitory synaptic currents or a specific voltage-gated ion channel, and so on. As always, we recommend studying the literature on your own preparation and your own question in detail. It is important to reflect why a particular ion or additive has been used—in addition to pure empirical values, traditions and sometimes a bit of "alchemy", there are often concrete reasons!

To prepare the solutions, you should have a magnetic stirrer, a pH meter, an osmometer and of course a laboratory scale. The water should come from a purified water system, be largely free of ions (as measured by conductivity) and, of course, free of any kind of impurities. Such water is available in most laboratories or institutes. You can also buy distilled water from a laboratory supply store, but if you use it regularly, it is more economical to make it yourself. All used substances or stock solutions should be kept carefully clean—this sounds trivial, but it happens again and again that in the heat of the moment spatulas or pipettes are put into several storage containers without having been cleaned. We have also seen scales in conditions that remind us of the kitchens of our college days… Finally, all glass containers should be free of residues, for which they are either washed in a laboratory dishwasher and dried in a drying cabinet or simply rinsed with distilled water and dried.

Some standard recipes for mammalian cells are discussed below to illustrate the basic principles.

Example

Extracellular solution for cultured or acutely dissociated cells:

140 mM NaCl
3 mM KCl
2 mM $MgCl_2$
2 mM $CaCl_2$
10 mM HEPES
20 mM Glucose ◄

The pH of this solution is titrated with NaOH to 7.35 (preferably on a stirrer, with sufficient mixing and waiting time). The solution approximates the composition of the extracellular space of mammals: the main ions are Na^+ and Cl^-, with concentrations between 130 and 150 mM being used. The concentration of K^+ varies in different formulations between about 2.5 and 5 mM, with higher concentrations leading to more positive membrane potentials and increased activity (e.g. in neuronal cultures). The ratio of Mg^{2+} to Ca^{2+} also affects excitability as well as specific functions such as the contractility of muscle cells or the synaptic activity in neuronal networks. Significantly higher concentrations of up to 5 mM $CaCl_2$ are sometimes used and are said to have a positive effect on the stability of the seal. High concentrations of magnesium tend to antagonize calcium influx and reduce excitability. Finally, HEPES is used as a pH buffer and glucose (usually 10–20 mM) as an energy source. The osmolarity of such a solution is about 290–300 mOsm.

Example

Extracellular solution for brain slice preparations (artificial cerebrospinal fluid, aCSF):

124 mM NaCl
3 mM KCl
1.6 mM $CaCl_2$

1.8 mM $MgSO_4$
10 mM Glucose
1.25 mM NaH_2PO_4
26 mM $NaHCO_3$
Gassing with Carbogen (95% O_2 and 5% CO_2) ◄

ACSF is used as a standard in experiments on complex brain preparations. With this solution, thick tissue slices (up to approx. 450 µm) can be kept alive for hours. This requires an intensive oxygen supply, which we ensure in all storage containers by means of fine-bubble aeration (filter candles, aeration stones for aquariums or also perforated tubes). Similar to pH buffering in vivo, the pH is automatically adjusted to 7.4 by CO_2 and bicarbonate ($NaHCO_3$). The calcium and magnesium concentrations approximate the ion ratios in the extracellular space of the brain and have been optimized based on experience.

► **Tip:** Many experimenters use a different solution for tissue preparation than the one in which they will ultimately record from their preparations. This is especially true for the highly sensitive brain tissue, whose extensively branched cells suffer multiple injuries during the production of slices or during dissociation. In such solutions, much of the Na^+ is replaced by other substances to avoid the massive sodium influx during injury with subsequent depolarization and swelling of the cells. N-methyl-D-glutamine (NMDG) is commonly used, but there are also other protocols based on sucrose, potassium gluconate, tris(hydroxymethyl)aminomethane (TRIS), or choline. We refer to the relevant literature, e.g. Ting et al. (2018) or Vandael et al. (2021).

Example

Intracellular solution for single-cell recordings (high chloride)

145 mM KCl
5 mM NaCl
2 mM $CaCl_2$
2 mM $MgCl_2$
4 mM EGTA
10 mM HEPES
pH 7.3 using KOH

This formulation (Schweizer et al. 2017) is based on a physiologically high K^+ concentration, but an artificially elevated Cl^- concentration. This corresponds to the extracellular value, resulting in a chloride reversal potential of approx. 0 mV – chloride currents will then usually occur as inward currents, and inhibitory postsynaptic potentials will be strongly depolarizing. For many applications, this simple and robust basic recipe is quite sufficient. The addition of a low Na^+ concentration approximates normal conditions and has proven to be stabilizing in many cells. The concentrations of calcium and magnesium vary widely in different protocols. It is important that the Ca^{2+} concentration is kept low and buffered, here with 4 mM EGTA (values up to 10 mM are common, depending on the nominal Ca^{2+} concentration). The free concentration of Ca^{2+} can be calculated with open-source programs, e.g. the *Maxchelator* by Chris Patton (Bers et al. 2010). A faster buffering is achieved using BAPTA, which in some applications is important to bind incoming Ca^{2+} at the membrane. The pH is usually adjusted to about 7.3 (sometimes 7.2), slightly more acidic than in the extracellular space. HEPES (2-(4-(2-hydroxyethyl)-1-piperazinyl)-ethanesulfonic acid) is used for this purpose, and the solution is titrated with KOH of course.

Example

Intracellular solution for recordings in neurons (low chloride):

144 mM K-gluconate
4 mM KCl
10 mM HEPES
4 mM Mg-ATP
0.3 MM Na-GTP
10 MM Na_2-phosphokreatine
pH 7.3 using KOH

In this formulation, gluconate is the main anion, so the Cl^- concentration remains low (4 mM). Alternatively, aspartate is also used. Under normal extracellular conditions, a Cl^- reversal potential of about -90 mV results; thus, the opening of chloride channels (e.g., at inhibitory synapses) can be measured as hyperpolarization (current clamp) or outward current (voltage clamp) and easily separated from excitatory postsynaptic potentials or currents. In the case of the solution cited here from Rozov et al. (2020), various additives (ATP, phosphocreatine, GTP) were also used, which will be discussed below.

▶ **Tip:** Replacing potassium with cesium in the intracellular solution largely blocks currents through the constitutively active potassium channels. Due to the increased membrane resistance, the cell becomes electrically "more compact", so that the ratio of series to membrane resistance and the voltage control in distant compartments improve (Sect. 5.2.4.2). In addition, the noise in current recordings is reduced, because contributions of potassium channels are eliminated. However, the cell can neither maintain its resting membrane potential, nor form normal action potentials, so measurements in the current clamp configuration are very limited.

4.4.5.1 Additives

We have often pointed out that the electrolyte solutions on both sides of the membrane make the situation of patch clamp recordings somewhat artificial. Important cellular functions are lost, and sometimes the stability of the cell suffers. Often, the intracellular solution is enriched with additives that are supposed to mitigate these effects. We list some of the classics here but point out that there are no patent solutions or fixed standards, and that you should rely on the specific literature for your own preparation, as well as on your own experiences.

- ATP: Used to maintain ATP- or energy-dependent processes; typically, 2–4 mM as magnesium or sodium salt.
- Phosphocreatine-salts: Serve as fast energy source for anaerobic glucose combustion; typical concentrations are up to about 10 mM.
- GTP: Supposed to maintain the function of GTP-binding proteins (i.e. G-protein-dependent signal cascades); typical concentrations are up to 0.5 mM as sodium salt.
- Glucose, cAMP or cGMP (especially in cells with HCN or CNG channels, which are modulated by cyclic nucleotides) and other substances are added depending on the preparation and the question.

References

Bers DM, Patton CW, Nuccitelli R (2010) A practical guide to the preparation of ca(2+) buffers. Methods Cell Biol 99:1–26

Chen CC, Cang C, Fenske S et al (2017) Patch-clamp technique to characterize ion channels in enlarged individual endolysosomes. Nat Protoc 12:1639–1658

Danker T, Braun F, Silbernagl N, Guenther E (2016) Catch and patch: A pipette-based approach for automating patch clamp that enables cell selection and fast compound application. Assay Drug Dev Technol 14:144–155

Dodt HU, Eder M, Schierloh A, Zieglgansberger W (2002) Infrared-guided laser stimulation of neurons in brain slices. Sci STKE 2002:pl2.

Dodt HU, Frick A, Kampe K, Zieglgansberger W (1998) Nmda and ampa receptors on neocortical neurons are differentially distributed. Eur J Neurosci 10:3351–3357

Dodt HU, Zieglgansberger W (1990) Visualizing unstained neurons in living brain slices by infrared dic-videomicroscopy. Brain Res 537:333–336

Haas HL, Schaerer B, Vosmansky M (1979) A simple perfusion chamber for the study of nervous tissue slices in vitro. J Neurosci Methods 1:323–325

Hájos N, Ellender TJ, Zemankovics R et al (2009) Maintaining network activity in submerged hippocampal slices: Importance of oxygen supply. Eur J Neurosci 29:319–327

Hamill OP, Marty A, Neher E, Sakmann B, Sigworth FJ (1981) Improved patch-clamp techniques for high-resolution current recording from cells and cell-free membrane patches. Pflugers Arch 391:85–100

Kantevari S, Matsuzaki M, Kanemoto Y, Kasai H, Ellis-Davies GC (2010) Two-color, two-photon uncaging of glutamate and gaba. Nat Methods 7:123–125

Maier N, Morris G, Johenning FW, Schmitz D (2009) An approach for reliably investigating hippocampal sharp wave-ripples in vitro. PLoS ONE 4:e6925

Rozov A, Rannap M, Lorenz F, Nasretdinov A, Draguhn A, Egorov AV (2020) Processing of hippocampal network activity in the receiver network of the medial entorhinal cortex layer v. J Neurosci 40:8413–8425

Schweizer PA, Darche FF, Ullrich ND et al (2017) Subtype-specific differentiation of cardiac pacemaker cell clusters from human induced pluripotent stem cells. Stem Cell Res Ther 8:229

Sylantyev S, Rusakov DA (2013) Sub-millisecond ligand probing of cell receptors with multiple solution exchange. Nat Protoc 8:1299–1306

Ting JT, Lee BR, Chong P, et al. (2018) Preparation of acute brain slices using an optimized n-methyl-d-glucamine protective recovery method. J Vis Exp (132):e53825

Vandael D, Okamoto Y, Borges-Merjane C, Vargas-Barroso V, Suter BA, Jonas P (2021) Subcellular patch-clamp techniques for single-bouton stimulation and simultaneous pre- and postsynaptic recording at cortical synapses. Nat Protoc 16:2947–2967

The Practice of Patch Clamp Experiments

5

In the previous chapters, we discussed the theoretical foundations and technical prerequisites of patch clamp experiments. Now we can really get started: Here, we describe the individual steps to create a patch clamp recording. What needs to be done in which order, how to create the different recording configurations, and what to pay special attention to?

5.1 Establishing a Patch Clamp Recording

At the beginning of the measurement, our preparation is in the recording chamber, which is filled with extracellular solution or is being perfused. The well-chloridized ground electrode (wire or pellet) is in the bath, and some pipettes are ready for use in a covered dish. After we have selected a suitable cell (or a tissue region for the blind-patch mode) under the microscope, we fill a pipette bubble-free with intracellular solution (Sects. 4.4.3 and 4.4.5), just enough that the silver wire dips a few millimeters into the solution. The silver wire should be "fiddled" into the front third of the pipette for this and must be sufficiently long and well chloridized. Then, we clamp the pipette in the holder (good fixation by the sealing rings is important!) and apply positive pressure to the tubing connected to the holder with a remotely connected syringe. If a manometer is connected, set it to about 50 mbar for cell culture preparations and to about 100 mbar for slices. Now, the pipette can be moved into the bath, normally by using a swiveling or tilting device and the coarse drive of the manipulator.

▶ **Tip:** Sometimes the pressure in the pipette holder drops by itself before seal formation. You will notice this because the counterpressure on the plunger of the syringe decreases or because the tissue in front of the

F. C. Roth et al., *Mastering the Patch Clamp Technique: A Practical Guide*, https://doi.org/10.1007/978-3-662-70745-6_5

pipette tip is not "blown away" by the outflow of liquid. This can be very annoying, but can almost always be remedied by simple measures:

- Look for and eliminate leaks in the syringe-hose-pipette holder path.
- Check the seals in the pipette holder and replace them if necessary.
- Check whether the syringe plunger moves too easily and releases the pressure by moving out.
- Use a three-way valve that can shut off the tubing to the pipette after pressure application.
- Install a manometer to monitor the tubing system.

5.1.1 Step 1: Compensation of Offset Potentials

Prior to seal formation, junction and electrode potentials must be balanced, as discussed in Chap. 3 (Neher 1995). Using standard patch clamp amplifiers, the voltage clamp mode is almost exclusively used for seal formation, which we will focus on here. If current clamp experiments are planned, one should also adjust the offset compensation in the current clamp mode. In voltage clamp mode, initially there is no command potential (it is either deactivated in the control software or set to 0 mV). In this situation, the amplifier should ideally display 0 pA current. However, this is often not the case because the real potential has an offset and is therefore not equal to 0 mV. This results in a current that is typically a few hundred picoamperes with the open pipette in the bath. When switching to a different type of pipette solution, the offset may change. Let's assume that the amplifier displays a value different from zero when the pipette is inserted into the bath, e.g. 500 pA. This offset can be "zeroed" manually (older models) or in the control program, i.e., the offset is eliminated and the potential of the open electrode in the bath is defined as 0 mV, at which the holding current is 0 pA. However, during the course of the experiment, changes in the offset may occur, so that the initial setting is no longer correct. This is especially true for liquid junction potentials (LJP), which disappear when the pipette is tightly sealed onto the membrane (Sect. 3.2.3) (Neher 1992). Other cases, where larger and larger offsets appear over time, cannot be explained by the recording solutions. This can be recognized by the fact that they also occur when the pipette and the bath are filled with the same solution. This phenomenon is often caused by wear of the chloride on the electrode wire or on the ground electrode (bath electrode) or by a change in its contact surface to the solution. This happens when too little pipette solution has been filled in or if it has not completely flowed into the tip. In the case of the bath electrode, there may be fluctuating offsets depending on the height of the "solution level" in the bath, especially if the electrode is not made waterproof (by using glue or tight heat shrink tubing) above the contact surface. The chloride layer on the bath electrode (pellet) and the electrode wire should then be checked and the components replaced or re-chloridized if in doubt. Finally, unwanted potential changes may also occur if the concentration of chloride ions in the extracellular solution is changed during the recording. This affects the ion transfer between the silver chloride electrode and the bath, creating a new offset potential. In experiments requiring changes in chloride concentrations, an agar bridge should be used for grounding, so that the environment of the silver wire remains constant (Sect. 4.4.4.1). Just to be clear: A nominally chloride-free solution largely abolishes the principle of charge transfer on chlorinated silver wires. Therefore, we never actually go below 4 mM chloride in the pipette solutions.

▶ **Tip**

After long recordings, the pipette opening can be cleared of cell membrane residues by (considerable!) overpressure and the current (voltage clamp) or the potential (current clamp) can be read out again without command voltage or current injection. The offset can now be a few millivolts (current clamp), i.e., different from the originally defined zero point. It is debatable whether all previously measured potential values should be corrected by this amount, since it is not known when the offset occurred during the experiment. If the potential value is critical for the experiment, then an exclusion criterion should better be defined: If the voltage values before and after the experiment differ by more than x mV, the cell is discarded (x can be, for example, 2 mV). This may be painful, but is unavoidable, because the potential values are uncertain and the data thus unreliable.

5.1.2 Step 2: Estimation of the Pipette Resistance

The pipette is now in the bath and the offset potentials are balanced. To determine the pipette resistance and to monitor it closely during seal formation, a constantly repeated test pulse should be applied, e.g., a rectangular change in the command voltage of -10 mV for 50 ms at a repetition rate of 10 Hz (Fig. 5.1). Since the current through the open pipette is expected to be quite large (several nanoamperes), the amplification in this situation is set to a low value, typically 1 mV/pA. The average current is approximately zero, except for the response to the negative command voltage pulse (test pulse), which each time evokes a corresponding rectangular current. The amplitude of this current response is determined by the resistance of the pipette. Therefore, the pipette resistance can be calculated according to $U = R \times I$ (Ohm's law).

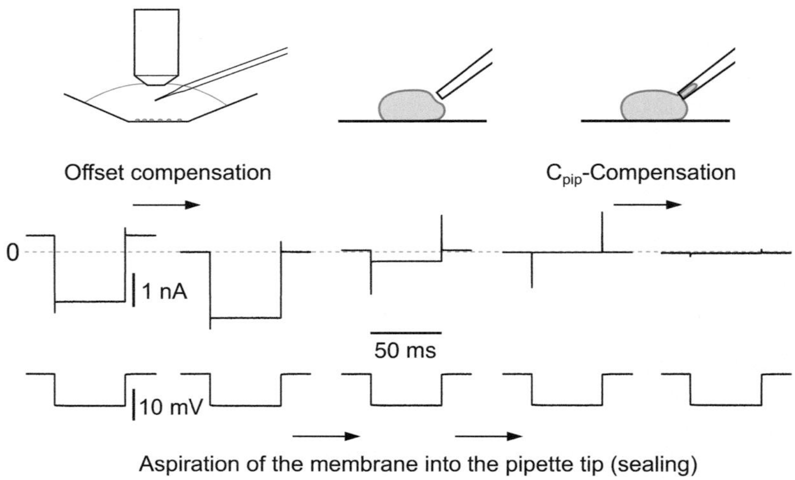

Fig. 5.1 Production of the cell-attached configuration (schematic drawing and typical currents)

Example

Amplitude of the test pulse 10 mV
Amplitude of the current response 2 nA
Resistance $R = 10\,mV/2nA = 10 \times 10^{-3}V/2 \times 10^{-9}A = 5 \times 10^{6}\Omega = 5\,M\Omega$ ◄

Many control programs for patch clamp amplifiers automatically calculate and display the resistance. Over time, one usually develops a feeling for the optimal range in which the resistance of a "good" pipette should be. Typical values are between 3 and 5 MΩ.

If you are using an amplifier offering bridge compensation in the current clamp mode, you can also set a comparable test pulse (10 Hz) of e.g. -300 pA and compensate for the pipette resistance until the response to the test pulse is no longer visible in the voltage trace. The compensated resistance can then be easily read in the control program or analogically at the amplifier. The bridge compensation is independent of the voltage clamp measurements described above and therefore provides a good control option in case of uncertainties in estimating the pipette resistance.

► **Tip**

If no rectangular current response to the command voltage pulse is found after immersing the pipette in the bath, the following errors may be present:

- Air bubble in the pipette.
- No connection between silver wire and pipette solution (e.g. insufficient filling).
- No connection between silver wire and pre-amplifier (e.g. loose solder joint).
- Ground electrode not immersed in the bath or not connected to the pre-amplifier.
- No connection between pre-amplifier and control unit.
- Incorrect program setting (e.g. voltage clamp mode with high command potential, so that the amplifier is out of range and generates a saturating current).

5.1.3 Step 3: The Formation of the Seal

Once the basic settings have been adjusted either manually or automatically, we are ready to approach the cell. To do this, one must first (except in the case of a blind-patch) locate the pipette tip under the microscope and make sure that the pipette is intact, free and clean, before bringing it close to the cell. To do this, the focus of the microscope is first moved towards the target cell and then the pipette is slowly and carefully moved towards the focal plane. Within a (brain) slice, it is important to move the pipette axially, i.e., in the direction of the pipette axis towards the cell, to minimize damage of the surrounding tissue. The correct starting position must be determined anew each time by moving forward, moving back, correcting the Y and Z position, and moving forward again. Meanwhile, you should always be careful not to "spear" other cells in the tissue and to keep enough overpressure on the pipette to clear the space immediately in front of the pipette mouth. It is important to get a feeling for the relative (vertical) position of the cell and the pipette by carefully "playing" with the focus. Especially with cultured cells (monolayers), the pipette can easily touch the bottom of the chamber and then inevitably break, which can be recognized by a sudden and sharp decrease in pipette resistance or the increasing amplitude of the current response to the test pulse. Just start over with a new pipette without getting unnecessarily upset—this has happened to all of us! In the best case, after some training, you should be able to find the pipette and the cell in the microscope's field of view and begin to form the seal between pipette mouth and the cell membrane.

In the case of isolated cells, the overpressure of the pipette should now be reduced to such an extent that the outflowing solution does not exert excessive forces on the cell or even blow it away. In slices, the cells usually tolerate higher pressures (up to 130 mbar). If possible, approach the cell surface obliquely from above along the direction of the pipette axis. Sometimes the surface of the cell can be slightly cleaned by the solution flowing out of the pipette before sealing, such that cell debris and extracellular matrix are removed. The immediate proximity of the cell membrane is indicated by two signals: 1) a slight decrease in the current response to the test pulse, as the resistance increases due to the membrane in front of the pipette, and 2) a characteristic indentation of the membrane (dimpling) caused by the solution flowing out of the pipette. As time goes by, you will develop a feel for your sample, and after a while, the manipulator movements, pressure reduction, observation of the cell membrane, and attention to the current amplitude will become automatic. In the immediate vicinity of the cell membrane, remove the overpressure from the pipette (by removing the syringe from the pressure tube or turning the three-way valve). Often this is enough to bring the membrane closer to the pipette tip and the resistance to rise steeply (i.e., the current response becomes very small). In most cases, however, gentle suction is needed. There are different "philosophies" about the timing and strength of the suction, ranging from "ramp-like and gentle" to "short and abrupt". We do not make a recommendation but encourage testing. However, coarse actions (sucking with maximum force) are definitely not useful.

During seal formation, the potential of the pipette should already be set close to the intracellular potential. Although not known, -60 or -70 mV is a good start value for normal mammalian cells. For many preparations, information can be found in the literature. When the resistance reaches about 30 MΩ (i.e., when the seal just begins to form), we often use an automated script to apply the negative command voltage. In our experience, this significantly promotes seal formation.

Once the seal is formed, the current trace is almost completely flat despite the ongoing test pulse, except for noticeable capacitive transients at the beginning and end of the test pulse (see below, step 4). The reason for the low current is the extremely high resistance that the membrane forms at the pipette tip (Fig. 5.1). With modern control programs, you can simply read the seal resistance on the monitor. Nevertheless, for a better understanding, we will go through an example.

Example

Our pipette with 5 MΩ resistance originally responded to the 10-mV test pulse with a current of 2 nA. After sealing, the trace appears flat on the monitor at low magnification. We now increase the gain to 100 mV/pA. The noise of the recorded current is now much more noticeable. With sufficient filtering (Sect. 5.3.2), the remaining current can be roughly measured and is about 2 pA in our example. According to Ohm's law, the resistance is 10 mV / 2 pA $= 10 \times 10^{-3}$ V / 2×10^{-12} A $= 5 \times 10^{-9}$ Ω, or 5 GΩ. This is a solid value for the seal resistance. The seal resistance is often even higher, so that even with high gain, the current response to the 10 mV pulse cannot be read accurately. ◄

What if the cells do not "bite"? There are many different recipes, tips, and legends for making "good seals". As always, our advice is to study the literature on your own preparation carefully, learn from successful colleagues, and bring enough patience for the necessary trial and error testing. We have already given some tips above. In case of persistent difficulties, we offer some further advice from our own experience below.

Tip

- The pipette is dirty. It should be stored dust-free and covered, unused of course, and filled with a clean, filtered solution. A highly contaminated bath solution can also clog the pipette mouth (e.g. by precipitating salts with incorrect pH or ion concentrations).
- The shape of the pipette is inappropriate. You have to try a few things—narrower or wider openings, thicker or thinner glasses, etc. Pipettes with a filament can sometimes cause problems (Sect. 4.4).
- The pipette is not sufficiently fixed and moves with pressure changes. Insert it completely into the holder, fix it well, and replace the seals if necessary!
- The micromanipulator is unstable. Check the assembly, care for consistent strain relief of all cables and tubing, and test the stability by long-term observation of the position of an open pipette in the bath (Sect. 4.2.2).
- The cells are particularly sensitive. Approach carefully, do not apply too much pressure, avoid excessive dimpling! Optimize the optics for a high-contrast image of the cell surface. If you have a choice: Large, round cells are the easiest to patch, very small or very flat cells are difficult.
- The cell or tissue is in poor condition and the membrane may be too stiff (no dimpling possible, the whole cell moves with the pipette). Choose optimal tissue, select "healthy", osmotically stable cells.
- The pressure tubing is kinked or leaky, so that after a short time there is no more overpressure or suction. If in doubt, dip the filled pipette into the bath, apply pressure and wait to see if the pressure holds.
- The pipette solution is not suitable (Ca^{2+}, Mg^{2+} concentration, pH, osmolarity, etc.). Check and possibly change it, based on the literature. Note: There are pipette solutions (e.g. those containing gluconate) where seal formation is more difficult and requires a bit more practice. Old pipette solutions can also cause problems. However, there is no strict rule—some prepare a new solution every 2 weeks, others are happy with preparing it once a year.

All electrophysiologists have experienced frustrating days, sometimes even weeks. Do not be discouraged, and do not try to force your way through. Before you start to "force" the pipette onto the cell or to suck the membrane into the pipette with too much force, you should go for a coffee, complain to a colleague and calm down. If no seals are obtained, even after systematic troubleshooting, the cells are often not in good condition. You should then optimize the preparation of the tissue or the culture conditions. Every effort is worth it!

5.1.4 Step 4: Capacitance Compensation

Once the seal is established, the pipette capacitance (C_p) should be compensated in voltage clamp mode. The capacitive charging currents of the pipette and pipette holder appear as "peaks" at the beginning and end of the test pulse (Fig. 5.1). The sampling rate of the signal must be high enough (usually well above 10 kHz) to accurately represent the short capacitive transients. These peaks in the signal are now minimized by injecting the charging current through a separate circuit outside of the current measurement. The theory of capacitance compensation was discussed in detail in Sect. 3.2.1. After manual or software-controlled compensation of the pipette capacitance, the capacitive artifacts are largely gone. The elimination of this non-biological signal also reduces the noise of the recorded current, which is of course a desired effect. However, the delaying effect of the capacitance on voltage changes and current measurements in the voltage clamp remains, since the pipette still has to be charged to a new potential as before—it is simply no longer visible in the recorded signal! A real acceleration of the charging processes can only be achieved by special amplifiers with overshooting current injection (supercharging). It is also essential to compensate for the pipette capacitance in the current clamp, where it has direct effects on the measurement accuracy. The time course of membrane potential changes can only be measured in real time with correct compensation (otherwise the capacitance acts as a low pass filter, i.e., it slows down all recorded signals). The pipette capacitance should be checked several times during longer recordings and adjusted if necessary. In general, it should be kept as low as possible.

Tip

- The solution film on the pipette should be kept as small as possible. To achieve this, the extracellular solution in the bath must be as shallow as possible so that the pipette dips in only minimally.
- In critical experiments, the pipette can be coated with water-repellent substances, e.g., by applying wax or Sylgard (Sect. 4.4.2).
- The pipette holder should be compact, dry, and clean.
- Finally, thick-walled pipettes reduce the capacitance. For low-noise recordings, glasses with the highest possible wall thickness should be used (Sect. 4.4.1).

When the seal is successfully formed, we are in the cell-attached configuration. You can now get started with your experiment or switch to one of the other configurations shown in Fig. 5.2 and discussed below.

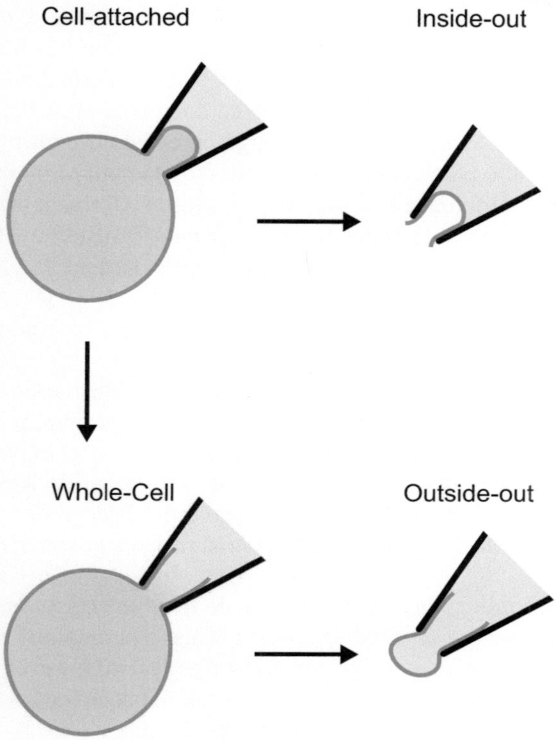

Fig. 5.2 The different configurations of patch clamp recordings

5.2 The Different Patch Clamp Configurations

5.2.1 The Cell-Attached Configuration

In the initially established cell-attached configuration (Fig. 5.2), the membrane under the pipette remains intact, but is deformed to a greater or lesser extent by the suction into the pipette, which can lead to activation of mechanosensitive ion channels. On the intracellular side of the membrane, all proteins are in their natural environment, i.e., all second messenger systems and the intracellular ion concentrations remain unaffected. This is a major advantage of the cell-attached arrangement over configurations where the pipette solution determines the intracellular environment (e.g. whole-cell). Many channels are modulated by the activity of phosphatases or kinases or by other intracellular mechanisms, which are then altered.

The cell's own resting membrane potential is also preserved in the cell-attached configuration. Only under the pipette, i.e. in a tiny section of the membrane, is the extracellular potential controlled by the amplifier. The resulting membrane potential at the patch is therefore equal to the difference between the (intracellular) membrane potential of the cell and the potential of the pipette, which can be determined by the control unit of the amplifier. With a "normal" resting potential between -60 and -80 mV and a voltage clamp command potential (U_{com}) also between -60 and -80 mV, there will not be a large potential difference between the inside of the cell and the pipette (in our example a maximum of 20 mV). Even if the electrical potential difference is zero, ions can still flow through the membrane, because the electrochemical gradient for ion currents depends on the distance between the membrane and equilibrium potential for the particular ion, which can be different from zero (Sect. 2.1).

Cell-attached recordings are used for several applications, for example:

- The observation of action potentials with a largely undisturbed intracellular milieu. Each action potential generates a small, transient current or a voltage fluctuation, that is visible in the voltage clamp or current clamp modes. This can be used to observe the "firing" of a neuron in its network. However, statements about the waveform of the action potential or observations of processes below the threshold for action potentials (subthreshold potentials) are hardly possible.
- Single-channel recordings in cell-attached mode show the behavior of the channels while preserving the intracellular milieu. This is particularly interesting if the channels can be modulated by intracellular signals. However, one has no control over the environment, i.e. the biophysical characterization of the channels is less complete than with isolated membrane patches.
- A particularly interesting application is the determination of the resting membrane potential of cells. This value is not as trivial to measure as some textbook chapters might suggest. If you pierce an intracellular measuring electrode through the membrane, the resting potential is likely to be altered by the injury. Even in the whole-cell configuration of the patch clamp technique (Sect. 5.2.4), artificial conditions prevail, because the intracellular milieu is altered by the pipette solution from the moment of membrane rupture. This is not the case in the cell-attached mode, which therefore allows measuring the "true" resting membrane potential. The prerequisite is that the membrane contains active potassium channels. A pipette solution is used that contains approximately the same concentration of potassium as the intracellular space, i.e., about 150 mM K^+. Since there is no significant concentration gradient, the driving force for potassium currents now depends only on the membrane potential of the patch. When the command potential of the pipette is systematically varied and currents through K^+-selective ion channels are measured, the current amplitude is exactly zero when the pipette potential is equal to the membrane potential of the cell (driving force is zero). By graphing the current amplitudes in a current–voltage plot and interpolation, one can determine the value of the noninvasively measured resting membrane potential quite accurately (Perkins 2006).

In cell-attached recordings, the sign convention is often a problem. When a negative potential is applied to the pipette in voltage clamp mode, the extracellular side of the membrane is locally hyperpolarized. This partially removes the negative intracellular membrane potential in the area of the pipette—in effect, the membrane below the pipette has been depolarized, i.e., the membrane potential has been shifted in a positive direction. Therefore, if you want to make the inside of the cell below the patch more negative compared to the extracellular side of the patch, you must set a positive command potential!

In the literature, there are different ways of presenting the potential—either the potential is given as specified by the amplifier (e.g., +60 mV), or it is taken into account that the intracellular potential in the area of the patch behaves exactly the opposite and in this case -60 mV is written. It is important to state clearly in the Methods section which convention is followed. The direction of the measured currents is generally viewed from the perspective of the cell. That is, a current of *positive ions into the cell* is an *inward current* and is conventionally given a negative sign and plotted *downward.* A current of *positive ions out of the cell* is an *outward current*, has a positive sign and is plotted *upward.* The opposite is true for *negative charges*, where the "technical current direction" is always given, e.g. a current of Cl^- into the cell is an outward current.

In cell-attached mode, the pipette is in the extracellular space, so for some measurements, the bath solution can simply be used as the pipette solution. In other situations, however, the pipette solution contains a special ion composition (e.g., in the resting membrane potential measurement example above) or additives, such as a transmitter that opens ligand-gated ion channels located under the pipette. Some example recipes for intra- and extracellular solutions are given in Sect. 4.4.5.

5.2.2 The Inside-Out Configuration

In the inside-out configuration (Fig. 5.2), the cytoplasmic surface of the membrane patch is exposed to the bath solution. To obtain an inside-out patch, a very good cell-attached seal should be established first. When the pipette is then slowly (!) pulled away from the cell, a piece of membrane will detach from the cell without the seal resistance decreasing noticeably. It is important that the pipette can move in a purely axial direction to avoid movements across the membrane. During the backward movement of the pipette, a slight negative potential should be applied (high voltage gradients damage the stability of the membrane), and it is recommended to reactivate the repetitive voltage pulse (test pulse) used during the seal formation to observe any possible deterioration of the seal resistance.

Sometimes the cell lifts off the bottom of the chamber when the pipette is withdrawn (especially with freshly dissociated cells) or a vesicle forms at the tip of the pipette. In this case, you can briefly lift the pipette out of the bath into the air. Usually, the vesicle or the cell bursts, and the desired inside-out patch is formed. The same effect can be achieved by dabbing the vesicle on an air bubble, a Sylgard ball, or a mineral oil droplet in the bath. However, briefly lifting the pipette into the air has the advantage of simultaneously reducing the solution film on the outside of the pipette. It should then be dipped back into the solution as little as possible to keep the noise low (Sect. 3.2.1). Vesicles form less often when working in calcium-free solution, but this is not the natural environment of the cells and should therefore not last too long. Fast withdrawal of the pipette during formation of the inside-out configuration can prevent vesicle formation, but often results in a poorer seal. The stability of inside-out patches is increased by replacing a large part (not all!) of the chloride ions in the bath solution with sulfate. However, this can cause considerable offset potentials, and an agar bridge should be used as the bath electrode (Sect. 4.4.4).

The inside-out configuration is cell-free, i.e., we now only have the isolated patch on the pipette. The intracellular milieu is replaced by the bath solution, so that modulating substances such as ATP, cAMP, IP_3, spermine, etc. are largely washed away. The absence of these molecules favors the so-called rundown effect, in which the measured currents become smaller and smaller as the experiment progresses. This behavior is very pronounced for some calcium channels and is difficult to control. There are all sorts of recipes to prevent rundown, but there is no magic bullet. If the addition of magnesium-ATP, phosphocreatine or other substances recommended in the literature does not help, the cell-attached method or a perforated patch (Sect. 6.2) may have to be used for longer recordings.

The considerations and conventions regarding signs remain the same as in the cell-attached configuration, since the outside of the membrane still faces the pipette. The flow of positive charges from the formerly intracellular side into the pipette is therefore an outward current (plotted upward) and vice versa. The sign of the potential is also reversed to the number displayed by the amplifier: If you want to get a relative potential of -60 mV on the inside of the membrane, you have to apply +60 mV to the pipette. The resting membrane potential of the cell is of course irrelevant in this cell-free configuration, and the reversal potentials of the various currents are now entirely determined by the experimenter by selecting the "extracellular" (pipette) and "intracellular" (bath) solution.

Inside-out measurements are typically used to systematically study the effect of intracellular modulators on ion channels. In this configuration, you can precisely specify and freely vary the solution on the intracellular side of the membrane. For example, the cAMP effect on HCN channels, which are physiologically modulated by cyclic nucleotides, can be studied. It should be noted, however, that the environment of the membrane is now highly artificial and contains neither proteins nor the intracellular low-molecular-weight substances.

5.2.3 The Outside-Out Configuration

Another cell-free configuration is the outside-out configuration (Fig. 5.2). Here, the former intracellular side of the membrane faces the inside of the pipette and the outside faces the bath. The conventions of voltage and current direction are again intuitive: A current from the bath into the pipette corresponds to an inward current (representation downward), a command potential of -60 mV means that -60 mV is applied to the inside of the membrane, as one would specify it as the membrane potential of an intact cell (which of course is no longer present here).

The outside-out configuration is established after initially being in the whole-cell configuration (Sect. 5.2.4). Since this mode differs in many ways from the cell-free inside-out and outside-out configurations, we will first describe the outside-out configuration. Outside-out patch formation begins with the cell-attached configuration, but then the membrane underneath the pipette is broken through, so that the inside of the pipette is directly connected to the intracellular space of the cell. Starting from this configuration (whole-cell), the pipette is now carefully withdrawn (as always, move it in the axial direction). This causes a piece of membrane to separate from the cell and to form a sort of half vesicle in front of the pipette opening. The outside of this piece of membrane now faces the bath solution (Fig. 5.2).

As with cell-attached and inside-out measurements, the very small membrane patch and the high seal resistance result in an extremely low-noise measurement, in which, for example, currents through individual ion channels can be resolved. In these three configurations, the high resolution of the patch clamp method is fully exploited. As with inside-out measurements, ion gradients and reversal potentials are determined by the choice of solutions. Now, however, the inside of the pipette is connected to the intracellular membrane surface, and the bath solution corresponds to the extracellular space. Thus, a negative command potential actually corresponds to a negative potential of the (former) cell interior. This configuration is particularly suitable for the study of ligand-gated ion channels, since agonists or modulators of the channels can easily be applied in defined concentrations to the outside-out patch. Since the whole-cell configuration is always established first, it is possible to combine whole-cell and outside-out recordings. Thus, cells can be characterized and whole-cell experiments can be conducted before "pulling" an outside-out patch.

▶ **Tip**

- In cell-free configurations, important intracellular modulators may be missing, causing the ion channels to change their functional properties or to show a rundown in their response amplitude (see above). This should always be kept in mind and useful hints should be taken from the literature—it may be necessary to add ATP, cAMP or other molecules to the pipette solution.
- If single-channel currents do not rise vertically but appear rounded, despite good adjustment of the amplifier, one either has channels under the edge of the pipette or a vesicle has formed instead of a cell-free patch. The latter is favored by, for example, increased calcium concentrations. In this case, check the solutions and try again. Of course, it is also possible that the current signal is unintentionally

filtered too strongly (low pass, Sect. 5.3.2). Strong noise indicates insufficient compensation of pipette capacitance (Sect. 3.2.1).
If the seal breaks when attempting to form an isolated patch, it may be for the following reasons:

- The pipette diameter is too large.
- The seal deteriorated before the membrane was pulled away or was instable to begin with.
- One of the solutions used is inappropriate. Possible causes: too large difference in inside/outside osmolarity, wrong pH, wrong calcium concentration, too old solutions, too low concentration of Mg^{2+} in the bath solution.
- The pipette was not moved slowly enough and not along the pipette axis.
- High (physiological) bath temperatures can decrease stability. Measurements at room temperature are usually easier to perform (if possible for the question at hand).

5.2.4 The Whole-Cell Configuration

Whole-cell recordings (Fig. 5.2) are probably the most commonly used method of the patch clamp technique today. It differs from the configurations discussed above in two major ways:

1. It records from the entire membrane of a cell, not just the area under the pipette tip.
2. Therefore, it has a much higher noise level than recordings from isolated patches.

Whole-cell recordings reflect all contributions to the conductivity of the cell membrane. In general, currents through individual ion channels cannot be detected in such recordings because the signal-to-noise ratio is too low. The large number of ion channels also has advantages: The superposition of many events leads to a kind of averaging, so that essential properties of the channels can be captured without time-consuming evaluation of thousands of single-channel events. In addition, the whole-cell configuration is naturally closer to the native behavior of cells than isolated patches. Voltage clamp and current clamp recordings can be used to study processes involving many different conductivities (e.g., action potentials) or even entire cell associations (e.g., synaptic currents in neuronal networks). Whole-cell recordings are therefore more similar to the conventional intracellular recordings with so-called "sharp" microelectrodes, but also show significant differences:

- Only very small leakage currents occur, which is why whole-cell recordings are tolerated by very small cells that would be destroyed by sharp intracellular electrodes.
- Due to the high seal resistance and the electronic circuit, the noise is significantly lower than in intracellular recordings. This allows the measurement of very small currents, depending on cell type and measurement conditions down to approx. 10 pA (rough estimate).
- Both voltage clamp and current clamp measurements are possible. In principle, this also applies to conventional sharp microelectrodes, but with these, the single-electrode voltage clamp technique is more complex and less frequently used.
- After seconds to minutes, the cytoplasm is exchanged with the pipette solution, resulting in a well-defined (but artificial) composition of the intracellular milieu.

To achieve the whole-cell configuration, a cell-attached patch is formed first (Fig. 5.2) and then the membrane area under the pipette is ruptured. This results in a current response as illustrated in Fig. 5.3a. Given that the lateral leak resistance (the "seal") is maintained, the measured current does not flow through the small patch under the pipette (which no longer exists), but through the entire remaining membrane surface of the cell.

The membrane is usually ruptured by applying suction via the pipette. This can be done either with the syringe attached to the pressure tubing or (preferably!) by gently or pulsatingly sucking on the connected mouthpiece. You have to approach each preparation individually, and it is strongly recommended to start very gently and with as little suction as possible. Your main guide is the electrical signal, while you sometimes can see on the video monitor how a part of the membrane is drawn into the pipette opening and finally breaks. Visible movements of the cytosol or the entire cell usually indicate a too rough treatment, which usually leads to low seal resistances and high series resistances. Alternatively, similar to sharp microelectrodes, the membrane can be broken by a very short, strong current pulse, which is integrated in some amplifiers as a so-called zap function (e.g., 1 V for 25 μs). Not all patch clamp amplifiers provide sufficiently short pulses, so the pulse may have to be generated by the software or an external pulse generator.

During the transition to the whole-cell configuration, the predetermined command voltage is important: During the formation of the cell-attached configuration, a negative holding potential is already set, which corresponds approximately to the expected resting membrane potential of the cell (e.g., -60 or -70 mV). This facilitates the formation of the seal, but is also important because otherwise the cell would be clamped to 0 mV immediately after the membrane is ruptured—such "voltage jumps" damage many cells.

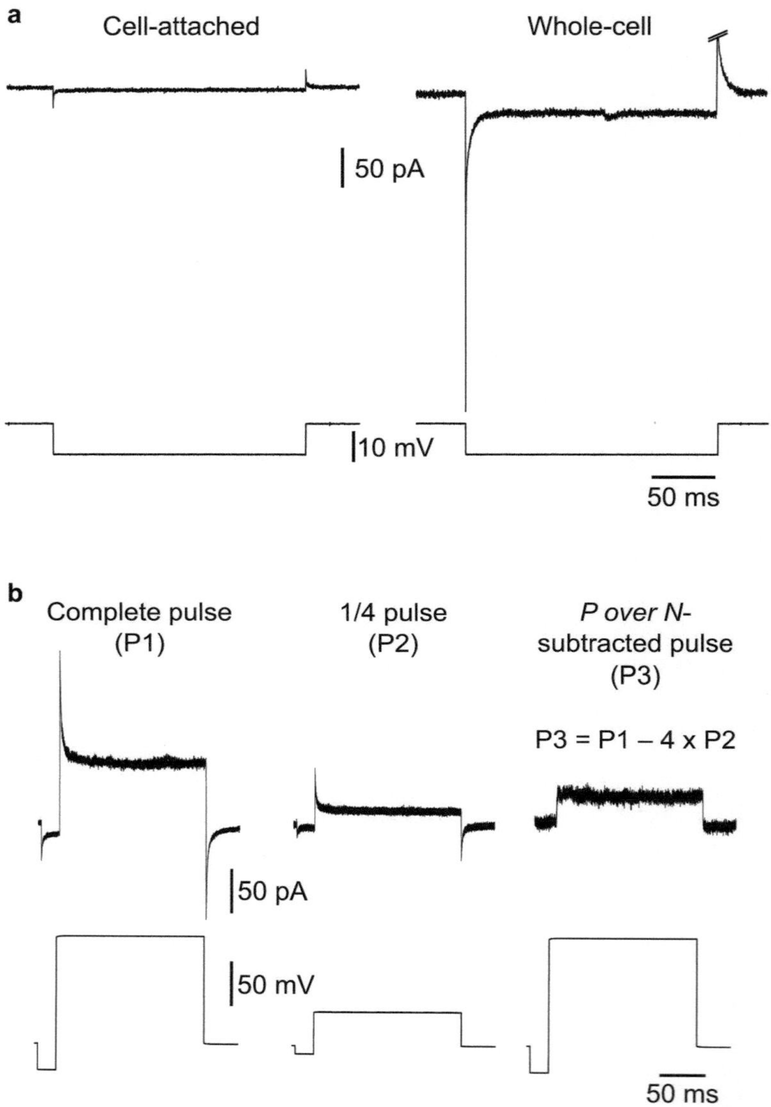

Fig. 5.3 **a** Current responses during the formation of a whole-cell recording from a granule cell in the dentate area of the mouse. In the whole-cell trace, the large capacitive current transient was curtailed in the positive direction. **b** Principle of P over N subtraction to eliminate capacitive artifacts and leak conductances. The outward current remaining after subtraction is a voltage-dependent potassium current

Once the membrane is ruptured, the current response to the test pulse changes, because instead of the patch, the entire cell membrane is now attached to the pipette, which of course has a higher conductance (lower resistance) and a larger area (greater capacitance). You can immediately see the large capacitive charging currents and, depending on the membrane resistance, also the "resistive" currents (based on the membrane conductance) that remain after the capacitive component has decayed (Fig. 5.3a). The recording can then be further optimized by trying to reduce the series resistance, and, depending on the recording protocol, by activating the compensations for capacitance and the remaining series resistance (Sect. 3.2).

The sign conventions are the same as for the outside-out configuration (the outside of the membrane is still facing outwards) and are also given as default by most programs and analysis routines: A negative command potential of the amplifier is associated with a negative membrane potential. If, in voltage clamp, positive charges leave the cell, the pipette will inject charges into the cell, which will be interpreted as an outward current (i.e. positive sign, plotted upward).

In whole-cell mode, the open connection between pipette and cytosol leads to an exchange of substances (cell dialysis), in which the inner environment of the cell adapts the pipette solution. This takes anywhere from seconds (for small, compact cells and low-molecular-weight ions) to many minutes (for branched cells such as neurons and for larger molecules) (Pusch and Neher 1988). Complete equilibration does not have to be achieved, e.g. in distal dendritic compartments or axons of neurons. These can at least partially maintain their internal environment despite the perfusion with pipette solution through enzymes and transporters, especially at elevated bath temperatures. The reverse exchange from the cell to the pipette is insignificant—the volume of the pipette solution is much larger than the cell volume and therefore its composition remains practically unchanged (exception: the targeted aspiration of the cytosol for mRNA analysis, Sect. 6.12).

The exchange of the cell interior with the pipette solution has both positive and negative aspects: On the one hand, the cytoplasm loses its natural composition and thus many factors that are important for the function of the channels under investigation. This can have a significant effect on the experiment, e.g. a rundown of channel activity as mentioned above. In such cases, one can switch to the perforated-patch method to avoid the loss of medium and small molecules (Sect. 6.2). On the other hand, it allows the control of the intracellular environment, to determine the reversal potentials of the ions of interest and thus to record under very defined conditions. Substances such as dyes can also be easily introduced into the cell via the pipette (Sect. 6.3).

5.2.4.1 Complicating Factors: Series Resistance and Capacitance

Section 3.2 discussed series resistance and the membrane capacitance in detail. It was also described how these factors are determined and how they are reduced by appropriate compensation circuits. Once a whole-cell recording has been established, this compensation must be performed in the voltage clamp mode in order to achieve the best possible voltage control and artifact-free measurements.

The compensation of the **membrane capacitance** (C_m) is done in the same way as described above for the (smaller) pipette capacitance, i.e. by current injection via a parallel circuit that does not contribute to the measured current. The magnitude of C_m depends on the total surface area of the cell (and thus its size) and is usually much larger than C_{pip} (in the order of several to hundreds of picofarads). The **series resistance** (R_s) is also compensated by injecting additional current, which can be variably adjusted. Since the time constant of the transient capacitive current is the product of $C_m \times R_s$, both values are adjusted simultaneously (iteratively) for optimal compensation until the capacitive current has been largely eliminated in the response to the test pulse (Fig. 3.3b). In very large or complex cells (e.g. neurons), however, this is not perfect because the remote cell compartments represent separate RC elements with their own electrical parameters, which cannot be compensated with a single compensation circuit (for theory, see Chap. 3). With older amplifiers, compensation is done manually using potentiometers, computer-controlled models provide automatic adjustment. At the same time, the control software of these devices returns the calculated values for R_s and C_m as important biophysical parameters.

Once you (or the software) have found an optimal setting for these two parameters, you can then adjust the strength of the R_S compensation between 0 and 100%. In fact, this circuit leads to a more precise and faster control of the membrane potential. However, the control loop contains positive feedback that increases the noise. If the compensation is set too high, it will even go into an uncontrolled oscillation, in which you will usually lose the cell. Realistically, in cells with a simple morphology, 60–70% compensation is achieved before the noise becomes too strong and disturbing.

While C_m is determined by the geometry of the cell and practically cannot be influenced, R_s can be reduced by some measures. Skeptics will argue that most of these tweaks are wishful thinking. However, we describe them anyway to appeal to your experimental and playful nature as a scientist.

▶ **Tip**

- Use large, low-resistance pipettes, if tolerated by the cells.
- Ensure sufficient osmolarity of the pipette solution (approx. 10 mOsm below the extracellular solution is a good guideline).
- Keep the Ca^{2+} in the pipette solution low by buffering (EGTA, BAPTA) to prevent resealing (i.e., the membrane re-closing under the pipette).
- Gently withdraw the pipette a little bit in the direction of its longitudinal axis after the whole-cell configuration has been established. If the membrane does not instantly open completely or tends to close again, further or repeated gentle suction may help. Sometimes, however, gentle 'blowing' or even a slight constant positive pressure is also useful.
- The error due to the series resistance depends on how large it is relative to the membrane resistance. The measurement can also be improved by increasing the membrane resistance. This can be achieved in many situations by using membrane-impermeable ions in the intracellular solution (e.g. replacing K^+ by Cs^+ or Cl^- (partially) by gluconate).

Series resistance may increase during an experiment. There are several reasons for this: Intracellular material can get into the opening, or the cell can shrink over time because the osmolarity of the pipette solution is too low. Sometimes, R_s can be reduced again by careful axial retraction of the pipette or equally careful suction through the mouthpiece. In many cases, however, a slight movement of the pipette (a mechanical drift), caused by the pipette holder or manipulator, can lead to the closing of the access. This can be recognized by the fact that the previous series resistance can be restored by repositioning the pipette. If this is not successful, the data are not comparable over the time course of the recording. In many publications, therefore, an exclusion criterion is given, e.g., that the cell is discarded if R_s has increased by more than 15–20% during the experiment. Such a rigorous approach hurts because you may lose a lot of data, but it is necessary. The remaining data set will be much more coherent and solid. After completing a series of measurements, you can also check the data to see if there is a correlation between certain parameters (e.g., the amplitude or the time course of currents) and R_s. This allows you to empirically determine to what extent the series resistance has affected the recording and at what value of R_s you should rather discard the measurement. We deliberately do not give a fixed upper limit for R_s in MΩ, as the values can vary greatly depending on the preparation, cell type and question. Finally, it should be noted that in many whole-cell recordings, compensation for R_s and C_m is omitted due to its complexity and to avoid the increase in noise. In addition, the R_s compensation makes it more difficult to monitor series resistance using a test pulse. You will have to weigh the higher accuracy of the measurement with compensation against the easier control of the stability without compensation and decide how to proceed depending on the question at hand.

P ove**r N subtraction:** Despite all compensations, capacitive current transients and leak currents always remain present during voltage steps and are mixed with the currents of actual interest. This is often disturbing when measuring voltage-activated conductances because the capacitive charging currents overlap with the current generated in the cell that you actually want to measure. Unspecific leak currents should react linearly to voltage changes and have a value of 0 at 0 mV. They therefore contribute to the measured current at all potentials different from zero and will cause errors in the analysis. You can estimate the leak current from the response to the known 10 mV test pulse and extrapolate it for larger voltage jumps. However, there is a special protocol available to remove capacitive transients and leak currents from the data by subtraction: Each voltage step in the protocol is applied several times in a miniaturized form, i.e., at 1/4 or 1/8 of the real amplitude. The responses to the small steps are then multiplied by 4 or 8, to reflect the leak component of the recordings at full amplitude. They are then subtracted from the"real" measurements (P over N subtraction, Fig. 5.3b). The goal is to measure the contaminating current transients separately, without the voltage-dependent activation of the channels of interest at full amplitude. To avoid activating channels, you can even reverse the protocol, i.e., perform hyperpolarizing 1/4- or 1/8-steps instead of depolarizing ones. At the end, you have a reduced (possibly inverted) image of the contaminating currents, which you can average or summate from several runs, invert again if necessary, then enlarge and subtract from the data trace. With this, the disturbing "spikes" of the capacitive artifacts and the leak currents are almost invisible. Of course, this "manipulation" of the raw data has to be mentioned in the Methods section, but it corresponds to the well-founded state of the art and is, in contrast to any arbitrary "beautification", a professional and reasonable procedure. Many recording softwares include this function and can perform it automatically. In this way, you can receive the subtracted signal already during the measurement.

5.2.4.2 Voltage Clamp in Large and Branched Cells: The Space-Clamp Problem

In Chap. 3, we already pointed out the difficulty of controlling the voltage everywhere within a large and/or branched cell, such as a neuron. A neuron can be imagined as a complexly branched circuit consisting of many small compartments, each with a capacitance and separated from the other compartments by a serial resistor. However, sitting at the soma with a single pipette, one cannot perform a separate R_s and C_m compensation for each of these compartments. Thus, in voltage clamp recordings, control over the local membrane potential in the periphery of the cell is poor (Armstrong and Gilly 1992).

Unfortunately, it is very difficult to estimate the extent of the space-clamp problem in individual cases. There is no universally accepted test of the quality of voltage control within a cell that is appropriate for every question. As a practical rule, keep the following in mind:

▶ The larger and more branched the cell, the worse the voltage control. Rapidly rising currents with large amplitudes are particularly difficult to control.

If you have the choice, you should prefer to record currents of small to medium amplitude in small, compact cells. If not (which is a realistic case), you should at least be aware of the problems that can occur. A clear indicator of insufficient R_s compensation and/or space-clamp problems are escape currents as shown in Fig. 5.4a. These artifacts are caused by the unintentional activation of voltage-gated sodium channels, indicating poor voltage control in the cell.

▶ **Tip**

- Escape currents often occur near potentials where voltage-gated ion channels are activated (Fig. 5.4). For Na^+ currents, this is approximately between -60 and -40 mV. It may be helpful to perform the measurement at slightly more negative holding potentials, e.g., at -80 mV instead of -60 mV.
- Sodium currents can be suppressed by adding the lidocaine derivative QX-314 (2–5 mM) to the pipette solution (Isaac and Wheal 1993).
- Of course, you can also add extracellular substances that block sodium currents. The most widely used substance is tetrodotoxin

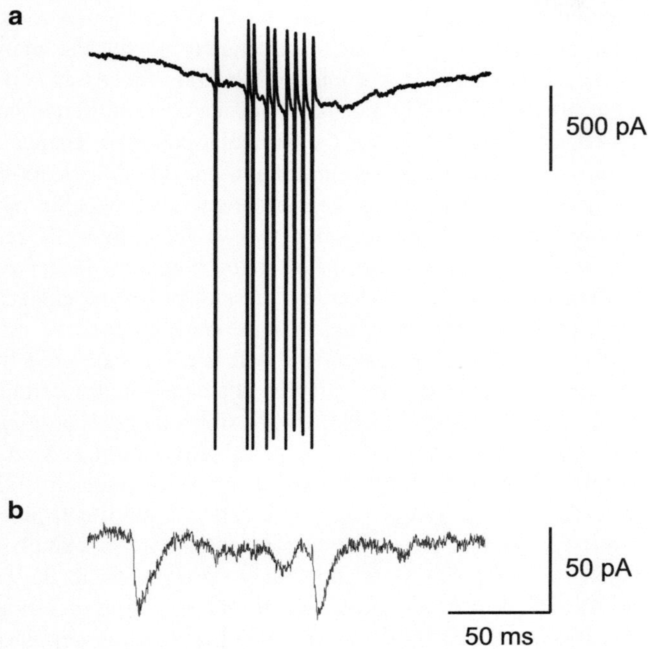

Fig. 5.4 Voltage clamp artifacts. **a** Escape currents in a recording from a retinal ganglion cell. The sharp inward currents are triggered by voltage-dependent sodium channels, which should not be activated at all with the formally applied voltage clamp. **b** Recording from the same cell type with reduced series resistance—here excitatory postsynaptic currents can be measured without the artifacts. Note the tenfold smaller calibration of the current amplitude. (Original data from Isabella Boccuni, Heidelberg)

(TTX), a toxin found in some pufferfish (fugu). However, extracellularly applied substances affect all cells in the preparation, whereas intracellular administration in the recorded cell leaves the neuronal network of a brain slice intact.

- Sometimes you are not interested in the entire cell, but only in the conductances at the soma. In these cases, you can create a nucleated patch containing practically only the somatic membrane and the cell nucleus it encloses. This preparation, which is ideal from the point of view of the space-clamp problem, is described in more detail in Sect. 6.4.

5.2.4.3 Current Clamp Measurements

In the whole-cell configuration, you can also perform recordings in current clamp mode (CC mode), in which you measure the membrane potential and can inject currents (Sect. 3.1.2). This configuration allows the electrical behavior of the membrane to be observed approximately as it occurs naturally: The membrane potential changes depending on the conductances that occur. Often, various ion channels are involved, so the method is not suitable for a detailed biophysical analysis of individual conductances. However, it is often used to study the actual cell physiology. By injecting current, you can test the behavior of the cell at different potentials and often obtain a characteristic profile for the particular cell type. Typical applications of CC recordings in the whole-cell configuration are the classification of neurons based on their intrinsic properties or the measurement of synaptic inputs and the resulting behavior of individual neurons in networks. In addition, CC recordings are used for the observation and modulation of action potentials in nerve, muscle, or sensory cells (including smooth and cardiac muscle) or for the characterization of spontaneously active cells (e.g., pacemakers in the sinus node).

In CC mode, the resting potential of the cell is often measured first (without current injection). Although this value may be characteristic for certain cell types, it is only of limited significance. First, the resting potential is often somewhat depolarized at the transition to the whole-cell configuration and must be stabilized by injecting negative current to obtain cell-friendly values of -60 to -80 mV. Secondly, from the first second of the whole-cell measurement, pipette solution flows into the cell, so that the membrane potential is increasingly determined by the artificial ionic conditions and no longer corresponds to the native value. Nevertheless, a short current-free measurement of the resting membrane potential is often useful, if only to identify damaged cells (permanently depolarized by leak currents) and discard them directly.

Subsequently, CC experiments can often be used to document the so-called intrinsic membrane properties of the cell by applying rectangular hyper- and depolarizing current steps. This is particularly interesting for excitable cells, whose action potentials are characteristic for the cell type in terms of shape and frequency. From small hyperpolarizing steps, the passive properties of the membrane (R_m and C_m as well as the so-called time constant) can be determined (Fig. 3.4 and Sect. 3.2.2.2).

Finally, it should be noted that the dSEVC amplifiers (Sect. 3.1.3) work on a completely different principle and are largely free of measurement errors due to R_s.

5.3 Hum, Noise, Filtering and Sampling

In this section, we will discuss possible interference signals and how to set the filters. We have already said a lot about this in Sect. 4.3.2, here we describe the practical procedure when disturbances occur.

5.3.1 Interference Signals/Noise

Mains hum is caused by electromagnetic induction due to the 50/60 Hz alternating voltage of the public power supply system, which is transmitted to the components of the rig via sockets, lamps and particularly by power supplies of equipment. We have discussed in detail in Sect. 4.2 how to protect the rig from such interference by grounding and shielding. Here is a checklist to be used in the event that a humming noise still occurs even though all the rules have been followed.

▶ **Tip**

- The electrical devices of a rig should be on a separate power strip, preferably on a separate circuit.
- When selecting equipment, give preference to direct current models whenever possible, especially in the case of stimulators with cables close to the pipette.
- Ground in a star shape over a collection point. Avoid loops/coiling of cables!
- Check the ground electrode for good conductivity.
- Avoid or eliminate moisture and salt residues in the electrode holder or around the bath.
- If the humming persists, remove all grounding and reattach cables one by one with an open pipette in the bath to observe each component's contribution. If in doubt, shield the chamber with a Faraday cage or conductive, grounded foil.
- Check tubing for bath perfusion and try different materials. Place pumping devices as far away or as well shielded as possible.
- Finally, a filter can be used to selectively remove the 50/60 Hz noise from the signal before recording. However, such measures should always be the last resort, after all controllable sources of interference have been optimized (see below).

High-frequency noise: We have already said a lot about this in Sect. 4.2. There are many biological and mechanical sources of noise, and not all of them can be eliminated. However, a few things can be done to optimize the signal-to-noise ratio:

▶ **Tip**

Reduce capacitance:

- Shallow bath solution and low immersion depth of the electrode.
- Clean, dry pipette holder (not homemade, as many plastics are extremely "noise-prone").
- Low fill level of the pipette.
- Use thick-walled pipette glass.
- For high-resolution recordings, thicken the wall with Sylgard or wax.

Further measures:

- Highest possible seal resistance.
- Use large, low-resistance pipettes, with a shape that tapers steeply toward to the tip (this concentrates most of the pipette resistance at the pipette tip).
- Do not set the series resistance compensation too high!

5.3.2 Filtering

Filtering of the signals takes place at the output of the amplifier, after the actual measurement (Fig. 5.5). You can filter the signal analogously before digital sampling, but this cannot be undone. For particularly sensitive measurements, we recommend recording the unfiltered raw data in addition to the filtered data. During the analysis, the raw data can still be digitally filtered, allowing to set the cutoff frequencies as you wish.

Relevant filters for patch clamp experiments are low-pass (eliminates high-frequency noise, while low-frequency components pass), high-pass (eliminates slow drift, while fast changes pass) and band-stop (eliminates certain frequencies, especially the 50/60 Hz hum). There are a few things to consider with all three filters:

Low-pass filter**:** The name tells you what this filter does: It allows the low-frequency parts of the signal to pass through and filters out the high-frequency components. This effectively reduces the noise of the current signal. You can and should try this on your own preparation by recording (or analyzing the data) with different filter settings. Reducing high-frequency noise is often desired, but there are also dangers. Fast signals, e.g. the activation of a voltage-dependent sodium current or the opening of individual ion channels, correspond to high frequencies (this can be recognized by decomposing the recorded signal into its frequency components using Fourier analysis). If the high-frequency components are now cut off by the low-pass filter, these signal components will be selectively lost, and the edges of the corresponding currents or potentials will be artificially slowed down. This also reduces the "peak" of the signal, i.e. the amplitude of the current. In addition, filters can lead to different phase shifts for the individual frequencies and signals, or temporal correlations can be distorted. For your own application, you should make preliminary tests with and without filtering as well as with

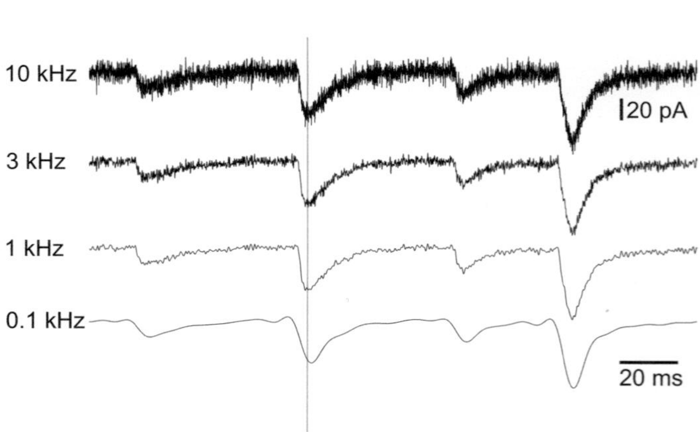

Fig. 5.5 Different filter frequencies and their effect on the signals. Excessive filtering (here 0.1 kHz) clearly distorts the time course of the synaptic currents

different filter settings. Only when these data (especially the kinetics) have been thoroughly analyzed, you should decide on a filter setting. It is also recommended to study the Methods sections of papers with similar questions. It is important to match the sampling rate of the data to the filtering (Nyquist criterion, see below).

High-pass filter: In addition to the low-pass filter, sometimes a high-pass filter is used, which cuts off low frequencies and lets high frequencies pass (again, the name says it all). The cutoff frequency can be at 1 Hz (1 s oscillation period), for example, to eliminate slow drift of the signal. Figuratively speaking, this keeps the average current on the zero line. This filtering, sometimes referred to as AC filtering, prevents the baseline of the signal from drifting out of range at high amplification factors.

▶ **Caution:** This is also an invasive procedure that can distort signals—for example, slow, long-lasting signal components can be transformed into seemingly transient decaying signals. Again, be sure to look carefully for differences between traces with and without filtering.

Band-stop: This variant usually appears as a selective filter against the 50/60-Hz mains hum that is to be selectively eliminated from the measured signal. Devices that selectively remove this interference are available on the market as Hum Bug. The technical principle is different from a real filter: The mains hum is recorded separately and subtracted from the original signal. This should not distort the actual measurement, even if it contains a real signal component in the range of 50/60 Hz. Even if it sounds a bit like a wagging finger, we repeat our advice here to convince yourself of the harmlessness of this intervention based on real data. With a dedicated search and grounding of the hum sources, you can often get by without such gimmicks.

▶ **Tip**

- In some experimental designs, similar signals can be repeatedly recorded and then averaged (e.g. postsynaptic currents, voltage-activated conductances). This drastically reduces unspecific high-frequency noise as well as annoying remainders of mains hum without active filtering.
- The signal should always be recorded at the highest possible amplification and then digitized. In this way, the high resolution of the patch clamp technique is best preserved in the data.
- Since modern A/D converters and computers can process and store very large amounts of data, it is often worth storing signals twice—once with and once without filtering. In this way, the comparison recommended above can be repeated at any time and, if in doubt, the original signal can be used, which can then be digitally filtered with different cutoff frequencies.

5.3.3 Digital Sampling

Another early stage of data processing related to filtering is the conversion to numerical values (digital sampling). Only digitized data can be processed by the computer, i.e., displayed graphically, used for the optimization of compensation parameters, or evaluated by analysis software. The measurement of currents or voltages is initially analog, and the devices used for discretization are called analog to digital converters (ADC). Conversely, there are also digital to analog converters (DAC) that convert numerical values generated by the computer into continuous, analog voltage signals, e.g., for controlling an electrical stimulator.

Some amplifiers have their own ADC (and DAC), or manufacturers offer them as an additional component together along with the control unit. In the case of modern digital patch clamp amplifiers, digitization already takes place in the preamplifier, i.e., immediately after the signal is generated. In other cases, the analog output of the amplifier is wired to the input of a separate ADC, to which further inputs and outputs can be coupled (e.g. field potentials recorded in parallel to the patch clamp experiment, or outputs for controlling the bath perfusion).

Two parameters determine the digitization: time resolution and amplitude resolution. The time resolution or sampling rate must correspond to the recorded signals. If you are dealing with fast physiological processes such as action potentials, you need to sample frequently to represent their time course (an action potential of 1–2 ms duration cannot be sampled with a frequency of 100 Hz, i.e., one sample per 10 ms). Formally, the lower limit of sampling is given by the **Nyquist criterion**: You must sample more than twice as fast as the fastest signal component. For an action potential of 1 ms duration, one could set the lower limit at 2 kHz (i.e., 0.5 ms per sample), but this would not reveal any details about the shape of the action potential. It would only provide the information that some voltage fluctuation occurred within 3 sample points. For any serious electrophysiological measurement, this would of course be far too low.

A more realistic approach is to look at the frequencies contained in the signal traces. According to the (extended) Fourier theorem, any signal can be represented as a sum of sinusoidal oscillations of different frequency and phase. For example, the steep rise of the action potential would be represented by very high-frequency sinusoids, so it is considered much faster than the 1 ms total duration. How fast one should digitize is often not easy to estimate. A low-pass filter can help: If you filter at 3 kHz, there should be little or no higher frequency components in the signal. Then, according to Nyquist, the lower limit of sampling would be 6 kHz (assuming you have convinced yourself that you can really afford to filter at 3 kHz, i.e. that the signal of interest is not distorted). Realistically, you would go well beyond this criterion and digitize at 50–100 kHz in our example with fast action potentials. Modern ADCs are powerful, and data can be stored in almost any quantity—so there is no need to spare sampling points here.

You have also to be aware of the signal amplitude. Modern ADCs work with an amplitude resolution of 16 bits, i.e., digital numbers with 16 digits. This corresponds to 2^{16} or 65,536 different values distributed over the possible input voltages of the ADC. For an input range of ± 10 V (or ± 10,000 mV), each possible digitization step would therefore correspond to 20,000/65,536 mV, or 305 µV. For the equally common input range of ± 5 V, it would be 153 µV. Fine differences, such as 100 µV, would not be detected with these settings. Therefore, you should amplify your analog signals sufficiently to make good use of the digitally detectable range. Otherwise, you will get rough values and step-like current or voltage traces in which the detailed shapes of the signals are lost and cannot be reconstructed later. At the same time, you should leave a small buffer, so that unexpectedly large currents are not amplified beyond the ± 5 or 10 V input range—if the ADC saturates, all values beyond that will become a smooth line. This is especially dangerous for whole-cell measurements with large amplitude fluctuations; a moderate amplification of 1 mV/pA is often used here. Modern amplifiers offer built-in variable amplification of the current signal from about 0.5 mV/pA up to 500, sometimes even 1000–2000 mV/pA. The voltage is usually amplified tenfold at the output of simple devices, but many amplifiers also offer variable gains up to 1000 times.

Example

We measure small currents through single channels in the range of 1 pA. With an amplification of 1 mV/pA, only 6–7 discrete digital numerical values would correspond to the amplitude of the currents—the data would appear vertically "pixelated" and precise statements about amplitude and noise would not be possible. Of the ± 5 V that the ADC can record, we would have used only 1/10,000! But if we feed the current signals into the ADC with 100 mV/pA, the amplitude of the current through an ion channel corresponds to 600–700 discrete points, which allows a decent resolution. Even better would be an amplification of e.g. 500 mV/pA. ◄

References

Armstrong CM, Gilly WF (1992) Access resistance and space clamp problems associated with whole-cell patch clamping. Methods Enzymol 207:100–122

Isaac JT, Wheal HV (1993) The local anaesthetic qx-314 enables enhanced whole-cell recordings of excitatory synaptic currents in rat hippocampal slices in vitro. Neurosci Lett 150:227–230

Neher E (1992) Correction for liquid junction potentials in patch clamp experiments. Methods Enzymol 207:123–131

Neher E (1995) Voltage offsets in patch-clamp experiments. In: Sakmann B, Neher E (eds) Single-channel recording. Springer, US, Boston, MA, pp 147–153

Perkins KL (2006) Cell-attached voltage-clamp and current-clamp recording and stimulation techniques in brain slices. J Neurosci Methods 154:1–18

Pusch M, Neher E (1988) Rates of diffusional exchange between small cells and a measuring patch pipette. Pflugers Arch 411:204–211

Special Applications 6

The basic principle of the patch clamp technique was soon adapted and extended in various ways after the first publications. Some of these variations allowed measuring completely new parameters (e.g., the capacitance of the membrane), others have made more complex preparations accessible (brain slices, in-vivo measurements) or optimized the technique for large series of measurements (automated patch machines). In this chapter, we will introduce the most important modifications to the patch clamp technique. The aim is to give a brief overview of their purpose and the technical principles – for details, please refer to the relevant literature.

6.1 Loose Patch

In the so-called loose-patch technique, the pipette is brought so close to the membrane that the resistance increases significantly (Marrero and Lemos 2007). However, no tight seal in the GΩ range is formed, as it would be the case in the cell-attached configuration. Only with the gigaseal can the sealing resistance between pipette and membrane be considered infinitely high, so that all currents must flow through the cell membrane and leak currents between pipette and membrane can be neglected. In the case of the loose patch, however, this approximation cannot be made, but a real voltage divider must be considered between the pipette and the membrane resistance on the one hand and the seal resistance on the other (Fig. 6.1a). In fact, it is possible that significantly more current flows through the leak towards the bath electrode than through the membrane patch, whose voltage the experimenter actually wants to control in voltage clamp mode. Ultimately, this results in an error of the type already known from the series resistance (Sect. 5.2.4). This error can similarly be reduced to a large extent in the loose-patch method by the R_s compensation, so that membrane conductances can

F. C. Roth et al., *Mastering the Patch Clamp Technique: A Practical Guide*, https://doi.org/10.1007/978-3-662-70745-6_6

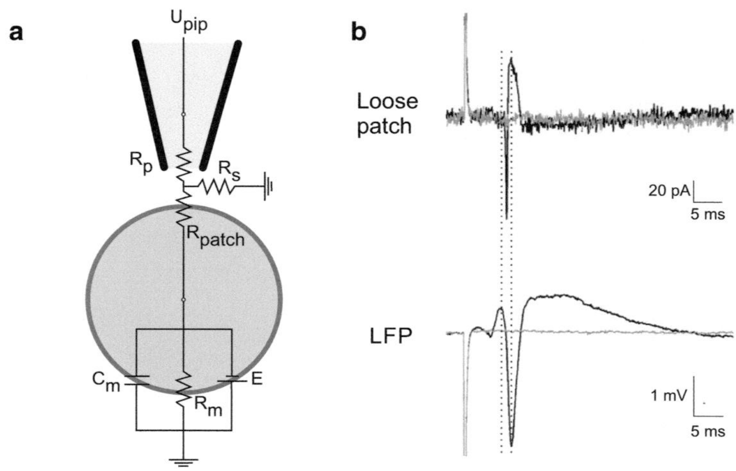

Fig. 6.1 Loose-patch clamp. **a** Principle of loose-patch recordings. There is a significant conductance between the pipette opening and the bath, referred to here as resistance R_s. **b** Recording from a pyramidal cell from a cultured mouse brain slice. After stimulation of afferent axons, "action currents" are generated in the loose patch recording (top), while a population spike is formed in a parallel recording of field potentials (LFP). (**a** Adapted from Anson and Roberts 2002. **b** From Daniel et al. 2013)

be determined quite accurately in voltage clamp mode without establishing a fixed and ultimately irreversible connection between the pipette and the membrane. However, this technique is much less suitable for precise biophysical characterization of voltage-dependent conductances than the conventional recording via gigaseal.

What is then the advantage of this less precise method? In some specimen, it is difficult to form a true seal because the extracellular matrix is too dense or too firm to allow good contact between the pipette and the membrane. To perform classical patch clamp experiments, these matrix proteins must be digested enzymatically or actively washed away, which can damage cells in sensitive tissues. The loose-patch method offers a less invasive alternative. In addition, it can be applied to moving cells, especially to muscle cells, whose surface can be literally scanned for activity without the need to immobilize them under unphysiological conditions. It is particularly helpful that the pipette can be moved and reused several times, in contrast to classical patch clamp experiments.

Technically, the loose-patch technique is initially easier to perform than a conventional patch clamp experiment. You approach the cell with an open pipette with or without visual control until the current response to the usual test pulse decreases, i.e. the resistance increases. Depending on the specimen, you will find out through testing which (ideally large) resistances can realistically be achieved without damaging the cell. Now you compensate for the "series resistance" (which is actually the sum of the series resistance and the "seal resistance") and record the membrane currents in this constellation in the voltage clamp configuration. This technique is closely related to the juxtacellular recording technique, which is used with various types of pipettes in current clamp or voltage clamp modes to non-invasively record action potentials of individual neurons (Pinault 1996). In fact, such approximations of glass pipettes to cells are very old, so that the loose-patch technique can be considered as a precursor of modern patch clamp technology (Chap. 1).

Loose-patch measurements can be performed with a wide range of pipette sizes. Standard patch pipettes with a diameter of a few μm and resistances in the MΩ range are used to record from subcellular membrane regions of individual cells. However, you can also deliberately pull large pipettes to study larger structures, such as axon bundles in a peripheral nerve or central fiber bundles. This gives you an averaged picture of the behavior of the cells, similar to the principle of field potential recordings, where you do not try to control the voltage (Sect. 2.2.2).

6.2 Perforated Patch

In the whole-cell configuration, the cell's internal milieu is gradually determined by the pipette solution, the cell is "dialyzed". As mentioned before, this creates a well-defined, but also artificial situation: The concentrations and equilibrium potentials of all ions are known, and you control pH, calcium buffering and osmolarity. For precise biophysical measurements, this situation is indeed ideal. It should be noted that large, highly branched cells (such as typical neurons) adopt the composition of the pipette solution in distal compartments (such as dendrites and axons) only slowly and often incompletely. This can lead to complicated situations with different ion ratios along the cell.

The specification of intracellular conditions is also problematic: The natural ionic milieu is destroyed, so that the native behavior of the cell can no longer be studied. In addition, numerous low-molecular substances are removed from the cell interior, e.g. the energy carrier ATP, but also reducing agents such as NADPH or glutathione and others. Many enzymes are inactivated or gradually lost – in the extreme case, one is dealing with a kind of membrane bag that has lost most of the cell-biological regulatory functions. This also affects membrane processes such as the function of primary or secondary active pumps or transporters and the phosphorylation state of ion channels. In addition, the natural protection against degradation processes is lost, allowing free radicals or residual proteolytic enzymes to damage the cell.

In practice, these processes have the following effects:

Some conductances show a time-dependent rundown, i.e. the channel function gradually decreases (typically over a few to several minutes). This is particularly pronounced in voltage-gated calcium channels, but also in some neurotransmitter receptors.

Other channels change their kinetics, so that even the most accurate biophysical characterization no longer reflects their native behavior.
The intracellular calcium concentration and dynamics are altered, so that all calcium-dependent processes no longer correspond to the natural state.
Virtually all second-messenger cascades (e.g. after activation of G-protein-coupled receptors) are interrupted.
Complex cell physiological processes such as exocytosis are interrupted or severely altered.

To avoid all this, numerous recipes for intracellular solutions have been developed (examples in Sect. 4.4.5). However, they do not change the fact that the situation is artificial. With the perforated-patch method, Horn and Marty (1988) introduced another, fundamental solution to the problem: access to the cell is not created by macroscopically breaking the membrane, but through small, selectively permeable pores. Such pore-forming proteins are produced by some bacteria, for example. When they are added to the pipette solution, they gradually migrate into the piece of membrane below the pipette, so that the recording slowly transitions from the cell-attached to the whole-cell configuration (Fig. 6.2). This can take many minutes, so that the series resistance decreases continuously over a long period of time until sufficient electrical access to the interior of the cell has been established. Time windows of 30 min or more to reach acceptable values of R_s are quite realistic, so the method requires very stable preparations. Ideally, series resistances of less than 10 MΩ are achievable, which is equivalent to a good classical whole-cell measurement. If the pore-forming substances are already present at the pipette tip at the beginning, they can cause low seal resistances, because they immediately start to make holes in the membrane. To avoid this, the pipette tip is usually first filled with a small amount of conventional internal solution (tipfilling; Sect. 4.4.3) and then backfilled with solution containing the pore-forming substances.

▶ The lipophilic pore-forming agents are poorly soluble in water. Therefore, stock solutions are usually prepared in solvents such as methanol or dimethyl sulfoxide (DMSO), at various concentrations suggested by the manufacturers. It is advisable to use as high a concentration as possible to minimize the volume of the toxic solvent added to the actual pipette solution. The solutions are considered unstable and should be prepared just before use.

The perforated-patch technique can also be performed with automated patch clamp systems (Sect. 6.10). Various pore-forming substances are used, which are briefly listed here:

Nystatin is a bacterial polyene macrolactone with a molecular weight of almost 1000 g/mol which forms circular holes in the membrane that are permeable to water and very small molecules such as urea. The channels are not or hardly permeable to divalent ions (Ca^{2+}, Mg^{2+}), but monovalent cations are well conducted

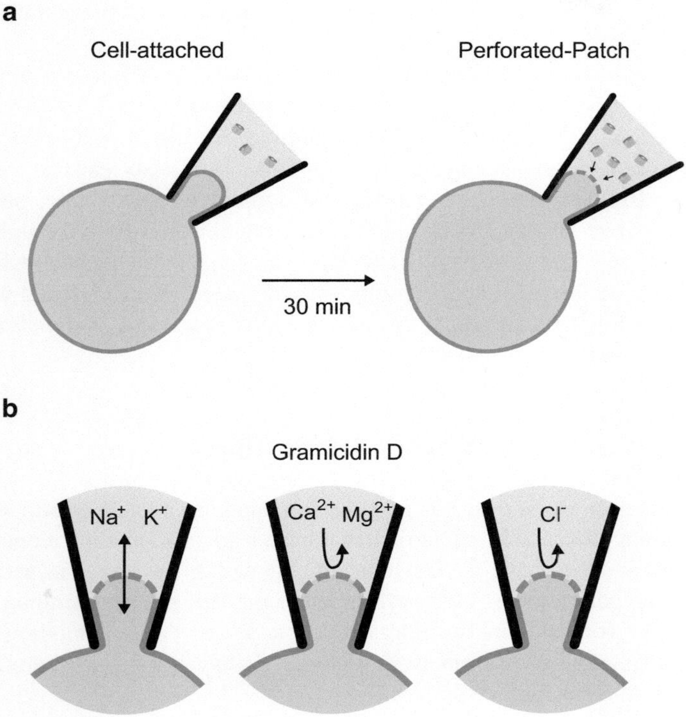

Fig. 6.2 Perforated patch. **a** Pore-forming agents migrate from the pipette to the membrane and are gradually incorporated. **b** Illustration of the permeability of different ions for gramicidin D

(e.g. sodium and potassium; cesium only when using an appropriate pipette solution). Chloride is also conducted, although the conductance values vary. In any case, it must be assumed that the chloride concentration in the pipette will significantly affect the intracellular concentration of Cl^-. This can lead to changes in the equilibrium potential for chloride, but also to cell swelling or offset potentials (Sect. 3.2.3). In such cases, it is often useful to work with pipette solutions containing a low chloride concentration (e.g. 10 mM) and to use large, impermeable anions (e.g. sulphonate, gluconate). Suggestions are given in Sect. 4.4.5.

Amphotericin B is very similar to nystatin in many respects, but it makes larger pores and thus produces lower series resistances.

Gramicidin D is a mixture of homologous, channel-forming peptides from *Bacillus brevis*. The special feature of gramicidin is its selective permeability for cations, i.e., chloride ions cannot pass through the channel. This allows patch clamp recordings to be performed in which the intracellular chloride concentration is not disturbed, so that, for example, the function of inhibitory synapses can be studied under realistic conditions. This is exactly the reason why the substance was introduced (Rhee et al. 1994). It is also not very water-soluble (DMSO stock solution), but due to the slow pore formation, it usually does not require separate tip filling.

β-Escin is a saponin, not a peptide, and is also suitable as a pore-forming agent, although it is used less frequently than the other substances. It is water soluble, so it is much easier to add to the pipette solution. β-escin is said to have slightly higher success rates in patching than the peptides, and also leads to good series resistances < 10 MΩ. The pores formed are apparently quite large, so that not only calcium, but also molecules up to small proteins can pass through. β-escin is especially used for the analysis of voltage-gated calcium channels, which show a greatly reduced loss of function (rundown) compared to the classical whole-cell configuration. Intracellular signal cascades (second messenger systems) also seem to be better preserved.

6.3 Intracellular Substance Application

For many research questions, it is of great advantage to be able to add substances from the intracellular side of the membrane or to change their concentration. Examples include exchanging the ionic milieu within the cell, the activation of second messenger cascades by cAMP or other mediators, and changing the buffering capacity for calcium by inorganic chelators or calcium-binding proteins. Typically, a baseline measurement is required, followed by a comparative measurement after the change.

This can be done directly in the inside-out configuration (Sect. 5.2.2), but only on an isolated membrane patch. What if you want to change ion concentrations, modulate the function of channels from the intracellular side, or change the capacity of calcium buffering in a whole-cell recording? In principle, the whole-cell method allows direct control of the intracellular milieu through the pipette solution. In larger and branched cells, the perfusion of each compartment may take many minutes, so you can sometimes use this time for a "before and after" comparison. However, you don't know exactly what the interior of the cell was like before, and with the ongoing exchange of the interior of the cell, it is impossible to get a clean separation between the initial situation and the subsequent effect of the substance. Ideally, you should first perfuse the cell with a solution (for a long enough time to establish an intracellular equilibrium), perform a control measurement and then add the substance. The following methods are available:

Multiple recordings from the same cell: If the pipette is carefully withdrawn from the cell in the axial direction after a whole-cell measurement, the membrane usually closes again and you get an outside-out patch (indicated by the very small current response to the −10 mV test pulse). It is then safe to assume that the cell membrane has closed as well. With a new pipette, you can re-establish a whole-cell configuration and do a second recording from the same cell with a different pipette solution. This procedure is surprisingly reliable with well-preserved cells.

Pipette perfusion**:** You can also change the pipette solution directly in the pipette by introducing a fine, flexible quartz glass capillary through an additional lateral access on the pipette holder (Tang et al. 1990; Aquila et al. 2015). This capillary is pushed far forward in the actual patch pipette and its contents can be released into the pipette solution with moderate pressure. At the same time, suction is applied through the usual pressure tubing (Fig. 6.3b), so that the cell is not exposed to any net positive pressure overall. The principle of a "pipette in the pipette" sounds a bit adventurous and does require some trial and error and finesse. However, after some early DIY solutions, commercially available devices have been developed that greatly simplify the process. Today's standard is a

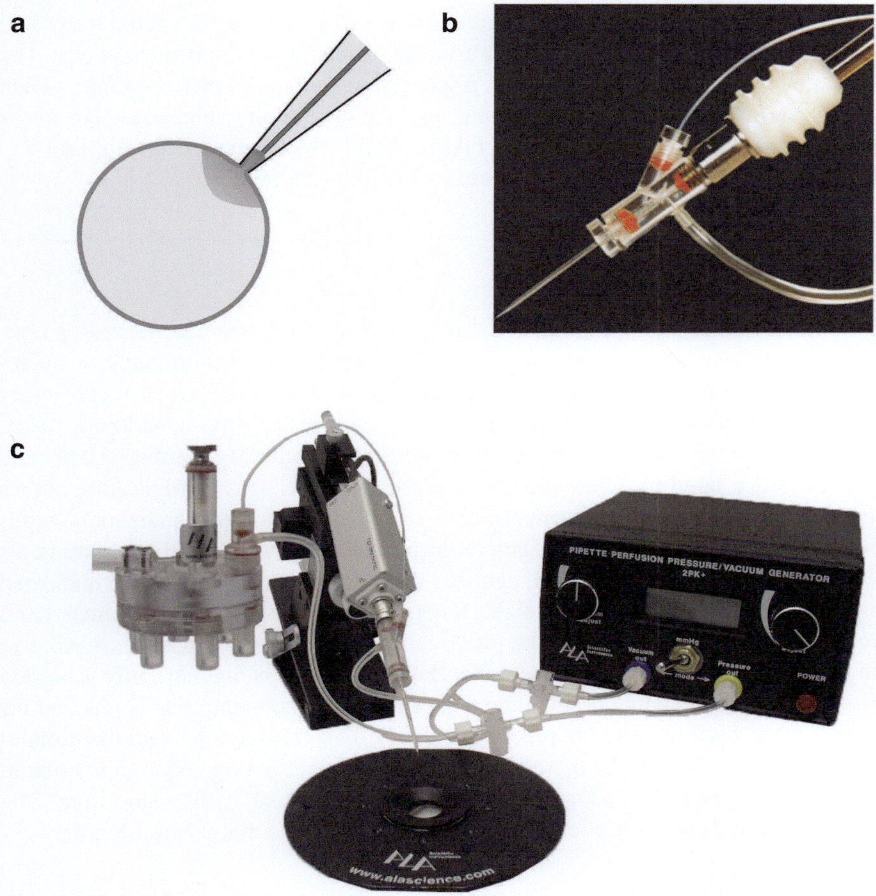

Fig. 6.3 Pipette perfusion. **a** Principle of intracellular substance application via the patch pipette. There is positive pressure on the small tubing inside the pipette and suction applied to the pipette itself. **b** Pipette holder with additional port for substance application. **c** System for pipette application. Positive and negative pressure are dosed with the control unit (right), on the left is the storage vessel for the solutions. (Image courtesy of ALA Scientific Instruments, USA/npi electronic, Tamm, Germany)

pipette holder with an additional input for the quartz capillary and a device for the precisely dosed application of pressure and suction. Pipette perfusion can also be used for isolated patches (outside-out or inside-out), where only suction (i.e. no positive pressure) is used.

Uncaging: An elegant solution for substance application is the use of light. The patch pipette can be used to bring so-called caged compounds into the cell. These are biologically active substances to which chemical groups are attached with a photolabile bond that inactivate them. High-energy light (usually in the UV range) is used to remove these groups, allowing the remaining molecule to exert its effect. For this purpose, xenon flash lamps can be used, which work with intense UV light flashes of about 1 ms duration, but also lasers up to a temporally and spatially highly precise release of substances using two-photon microscopy. The fast activation of the substances is an important advantage of the technique, which is by no means limited to intracellular applications! Many substances are available as caged compounds, e.g., ATP, cAMP, IP3, NO, and most neurotransmitters. Similarly, you can release calcium from photolabile chelators.

6.4 Nucleated Patch

In some cases, an attempt is made to create a kind of intermediate recording configuration between whole-cell and outside-out, called a nucleated patch or nucleated macropatch. Here, the nucleus of the cell is used to detach a much larger membrane "sphere" from the cell membrane compared to the outside-out patch. Thus, the pipette is connected only to a part of the soma of a formerly branched cell (usually a neuron), while the processes are torn off. This configuration allows the study of macroscopic currents (i.e., the activation of many ion channels) without having to deal with the space-clamp problems of whole-cell measurements (Martina et al. 1998). In addition, agonists can be applied very rapidly to nucleated patches, allowing the analysis of the kinetics of ligand-gated ion channels (Zhu and Vicini 1997). In principle, the production of the nucleated patch is similar to an outside-out configuration, except that the cell nucleus is first carefully aspirated onto the pipette opening and then the adjacent part of the membrane is pinched off from the cell by retracting the pipette (Fig. 6.4). This results in a small membrane-enclosed sphere of somatic membrane plus nucleus that is large enough to contain many ion channels and allows very good voltage control at the same time. The technique requires good optics and some fine-motor skills, but it is quite reliable.

6.5 Recording from Subcellular Structures

Neurons, as mentioned above, are typically highly branched cells, whose electrical properties and activity can vary between dendrites, soma, and axon. A branched dendritic tree often involves complex processing steps with active and passive forwarding and modulation of input signals (dendritic integration). The axon, on

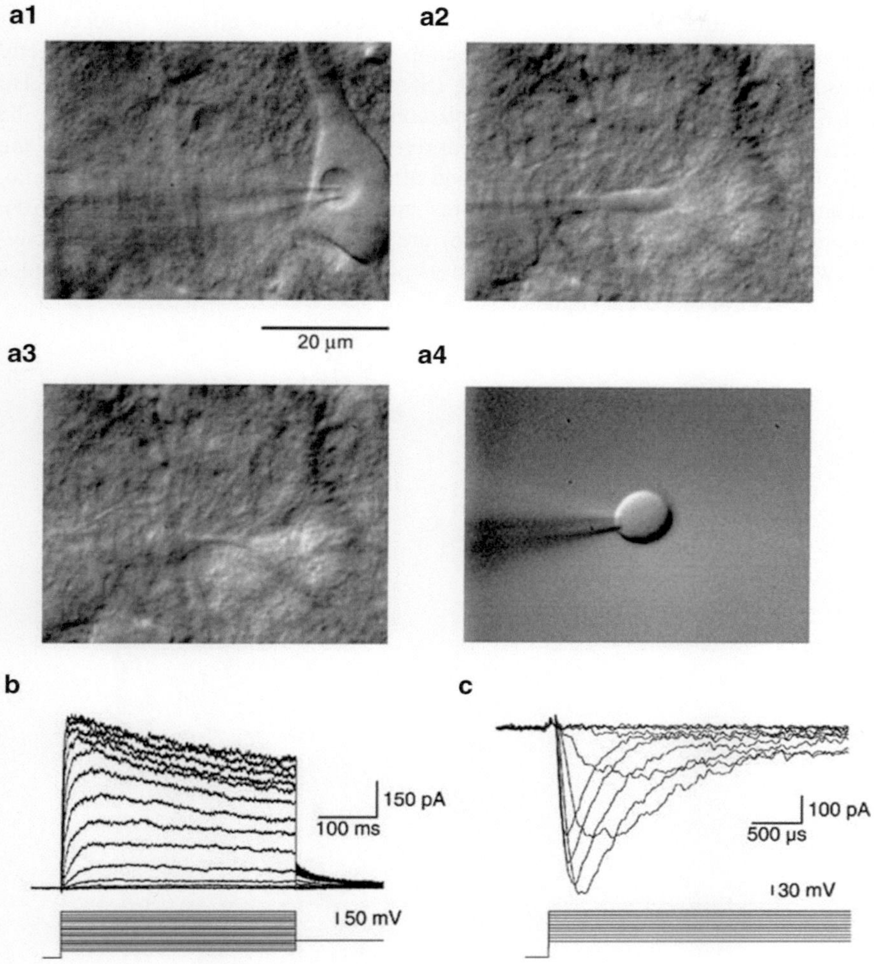

Fig. 6.4 Nucleated patch. **a** Preparation of a nucleated patch of a neocortical pyramidal cell in a brain slice preparation. **a1** Approach to the cell. **a2** Withdrawal of the pipette in the whole-cell configuration. **a3** Closure of the membrane between pipette and cell. **a4** Free nucleated patch. **b** Voltage-dependent potassium currents (voltage clamp). **c** Voltage-dependent sodium currents. These currents would not be measurable with sufficient voltage control in the complex, extended cell. (From Gurkiewicz and Korngreen 2006)

the other hand, features a special type of signal processing, the highly effective generation of action potentials, especially in the initial part near the soma (axon initial segment, AIS). The underlying mechanisms involve not only the spatial separation (compartments) and characteristic morphologies, but also the asymmetric distribution of receptors and ion channels, which lead to very different electrical properties.

For a long time, these local subcellular processes were difficult to access, but thanks to the optimization of patch clamp methods, they can now be directly measured with some effort and training (Stuart et al. 1993; Davie et al. 2006; Hu and Shu 2012): Usually, the whole-cell configuration is first established at the soma of a neuron, allowing a fluorescent dye contained in the pipette to enter the cell. The dye diffuses within the cell and after a relatively short time (5–30 min, depending on their diameter), dendrites and axons become visible (Fig. 6.5). Fluorescent dyes (e.g., alexa-hydrazide) are used that do not disturb cell activity (while using as little excitation light as possible) and thus allow simultaneous measurement of electrical signals.

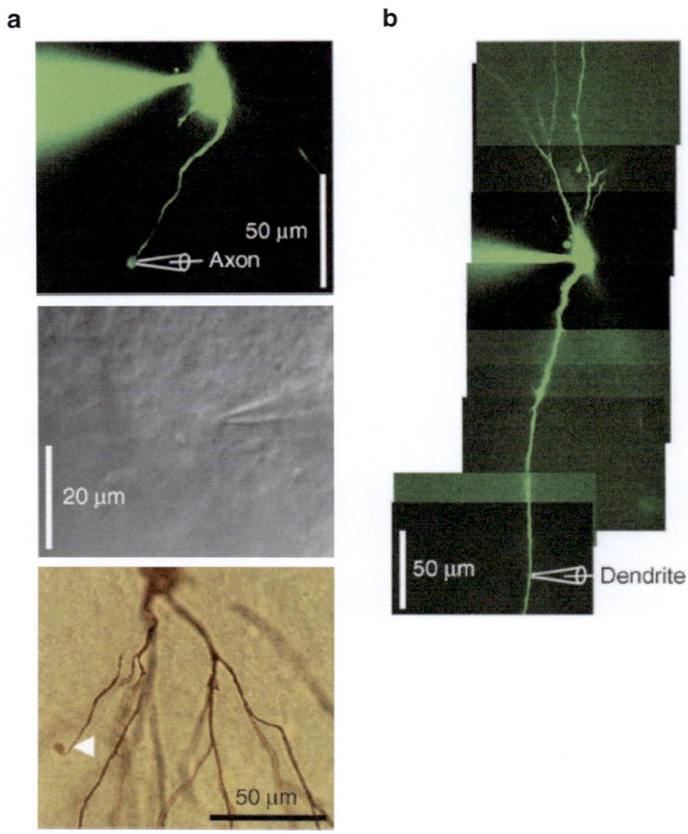

Fig. 6.5 Recordings from subcellular structures. **a** Simultaneous recording of soma and axon. Fluorescence image of the applied dye through the somatic pipette (top); IR-DIC photo of the axon bleb with pipette (middle); biocytin staining of the recorded cell for subsequent evaluation of the morphology (bottom). **b** Fluorescence image of a somatic and dendritic measurement. (From Kim et al. 2012)

An additional patch clamp configuration is then established with a second pipette (without dye), whose opening size is adapted to the respective target structure (dendrite approx. 8–12 MΩ, axon 12–20 MΩ). The fluorescent signal of the intracellular dye allows the desired subcellular structure to be found in the IR contrast illumination at high magnification and resolution (e.g. through overlay of the images). The gigaseal is then formed with the second small pipette. Of course, this type of patching requires a high degree of stability and very good optics. High-precision, drift-free manipulators and pipette holders are highly recommended: Drift movements disturb the whole-cell configuration in very small structures much more, as these tend to instantly block the access to the cell.

Once a cell-attached or whole-cell measurement has been established, the electrical connection between the "patched" structures can be tested using current pulses in current clamp mode: A current injection at the soma should (depending on the distance and length constant) cause a clearly measurable potential response in the dendrite or axon. Alternatively, one can use a second dye with a different excitation/emission profile for the subcellular structure and observe the superposition of the two fluorescence signals. However, this is unnecessarily expensive and complex for routine experiments. Once a subcellular structure connected to the soma has been successfully patched, one can measure the local signals during different stimulations or activity patterns and, for example, analyze signal propagation to the soma or from the soma to the peripheral structure. After establishing a whole-cell recording on subcellular structures, it is also possible to obtain outside-out patches to study the local ion channels and receptors.

6.6 Capacitance Measurements

In Sect. 3.2, we have already made estimates of membrane capacitance using simple, rectangular voltage jumps, and we have discussed their compensation. However, there are research questions where one wants to measure the capacitance of cell membranes with the highest possible amplitude and time resolution. The most important application for this is the exocytosis of vesicles, which leads to an enlargement of the membrane surface and thus of the capacitance.

Erwin Neher started very early after the fundamental publication of the patch clamp method with real-time measurements of vesicle fusions and introduced, together with Alain Marty, the basic method for time-resolved capacitance measurements (Neher and Marty 1982) (Fig. 6.6). They took advantage of the physical principle of the temporal change of electrical signals by capacitances: A sinusoidal input voltage generates sinusoidal currents at a capacitor with the same frequency, but with a phase shift of $1/2\pi$, i.e. a quarter of the oscillation period (cycle). This can be understood from the definition equation of capacitance:

$$I = C \times dU/dt$$

The capacitive current is therefore maximum when the change (time derivative) of the voltage is maximum, and zero when the voltage remains unchanged. A sine

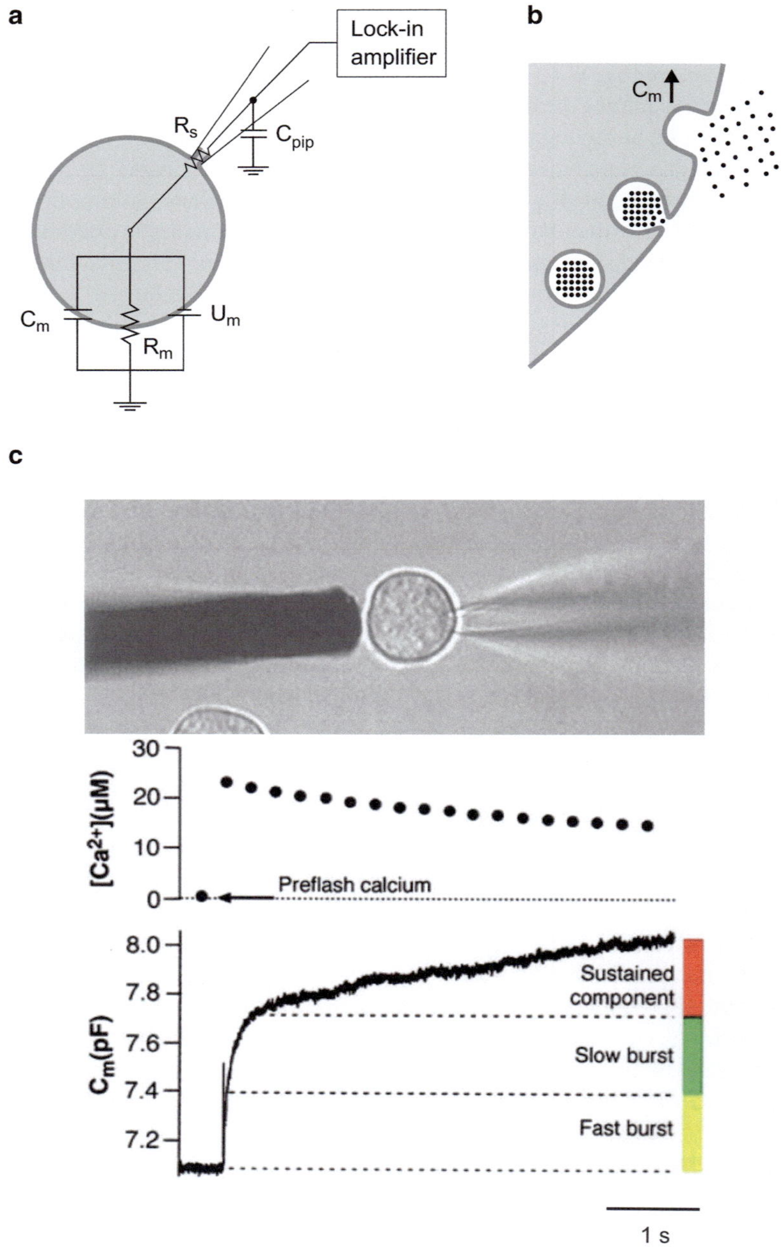

Fig. 6.6 Capacitance measurement. **a** Principle of the measurement. The target parameter is the capacitance of the cell membrane. **b** The fusion of vesicles causes an increase in the membrane area, so that the capacitance increases. **c** Original measurement of a chromaffin cell of the adrenal medulla. The patch pipette (whole-cell) is seen on the right, a carbon fiber is positioned on the left, which measures released catecholamines (adrenaline, noradrenaline) by means of amperometry. Calcium was released intracellularly by photolysis (uncaging; Sect. 6.3), followed by an increase in membrane capacitance in various phases. (**c** From Neher 2006)

waveform for U(t) produces a cosine waveform for I(t) that is shifted by exactly 90° or $1/2\pi$. For the "resistive" current component, i.e., the currents flowing through the membrane resistance, there is no such time or phase shift:

$$I = U/R$$

This allows the resistive and capacitive components to be well separated: If you look at the currents that oscillate exactly with the input signal, you get a signal that is proportional to 1/R, i.e. to the membrane conductance. If you look at the currents that are exactly shifted by $1/2\pi$, these are a measure of the membrane capacitance. This is achieved by so-called lock-in amplifiers, which multiply the sinusoidal signal of the command voltage with the signal of the current–voltage converter of the voltage clamp measurement. This product is maximum, when both signals are "in phase", i.e. oscillating synchronously (at $1/2\pi$). At this phase shift, the resistive component is zero, because it occurs exactly at the zero crossing of the command voltage. For those who are particularly interested, we refer to the calculation of capacitive and resistive currents using complex numbers, from which the phase shift results directly.

The signal of the lock-in amplifier is then low-pass filtered and provides a direct measure of changes in membrane capacitance (as well as a phase-shifted signal for changes in membrane conductance). For calibration, small changes in the capacitance compensation are used, which are set as accurately as possible beforehand. A complication to the simple measurement principle described above is that we are not only dealing with the membrane resistance, but also with the series resistance. This makes the above equations somewhat more complicated and must be taken into account, especially since all parameters can change, in particular the membrane conductance (e.g. when a calcium influx is deliberately activated by voltage jumps, which then induces the exocytosis of vesicles). We omit details here and refer to the literature at the end of the chapter (Gillis 2000; Lindau 2012). The same applies to a slightly modified procedure, the *Sine + DC* technique, which also measures and considers the absolute holding current (i.e. the current resulting from the difference between the membrane potential and the mixed equilibrium potential of all conductances). This method is somewhat more accurate, but also more complex and is therefore less frequently used. There are many other variations of input signals used (e.g. superimposed sinusoids of different frequencies, rectangular pulses) and combined with exocytosis stimuli (e.g., in addition to voltage jumps, the elegant method of photolysis of caged calcium, which does not induce a significant change in membrane conductance).

Some models of modern patch clamp amplifiers already offer phase-coupled signal analysis in their software, so that capacitance measurements can be performed directly. This should be taken into account when choosing an amplifier. The method is particularly sensitive with small, very high-resistance (i.e., low-conductance) and spherical cells. When measuring such cells in the whole-cell configuration, sinusoidal frequencies of up to 3000 Hz can be set providing a correspondingly good time resolution. For isolated "patches", especially in cell-attached mode, one can even go up to 20 kHz and thus observe the fusion of individual vesicles in real time.

6.7 Dynamic Clamp

We described the principle of the voltage clamp in Sect. 3.1.1, where a membrane potential is set and stabilized by appropriate correction currents. These correction currents are then a measure of the conductance of the membrane. A modification of this principle is the dynamic-clamp technique. As with the voltage clamp, it uses Ohm's law, which relates membrane currents, resistance (or its reciprocal, the conductance) and driving voltage (the distance of the instantaneous membrane potential from the equilibrium potential for the given conductance):

$$(U_m - E) = R_m \times I$$

or

$$G_m \times (U_m - E) = I, \text{ where the membrane conductance is } G_m = 1/R_m.$$

This principle can be used to artificially create or simulate membrane conductances (Fig. 6.7). Suppose we want to study the effect of a particular synapse on a neuron, say an inhibitory GABAergic synapse, which conducts chloride ions with a defined amplitude and time course. We know roughly what G(t) looks like – in our case a steeply rising and exponentially declining increase in conductance for chloride ions. In the dynamic-clamp procedure, this course of G(t) serves as a

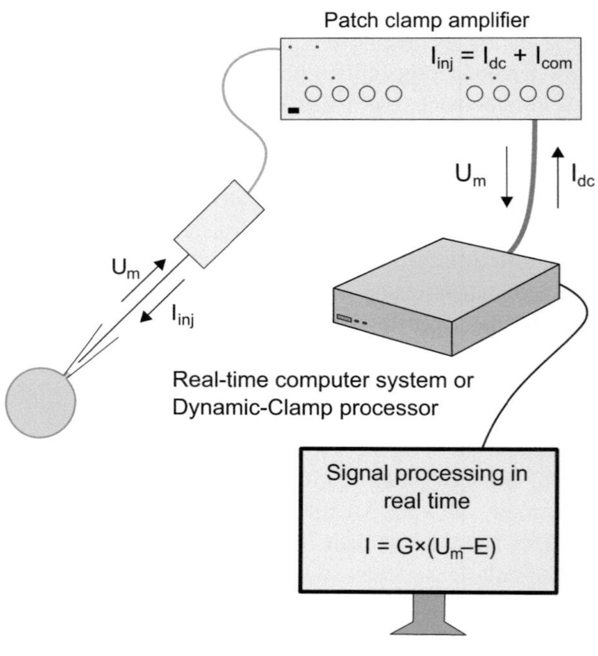

Fig. 6.7 The dynamic clamp technique requires feedback between the patch clamp amplifier and a dynamic clamp processor or software in real time

preset ("model") for our recording. We also know the equilibrium potential E for chloride ions, especially when the intra- and extracellular chloride concentrations are known in the whole-cell measurement. We can thus measure the instantaneous membrane potential U_m, so that we can easily calculate the current I, which would be generated by a GABAergic synapse, and inject it into the cell. As a result, U_m will change, as will our assumed synaptic conductance, which changes over time according to our pre-defined model. Using the model, we can then calculate a new instantaneous current I etc. In this way, we induce a change in U_m over time, which corresponds to an inhibitory postsynaptic potential (IPSP). If U_m looks like a "real" IPSP in the trace, our model for G(t) was plausible – otherwise the experiment has to be repeated with adjusted parameters for G(t). Ultimately, we are simulating a change in membrane conductance, testing its effect on the cell and in this way learning more about important membrane processes. Using the same principle, dynamic-clamp may also be used to "turn off" conductances by injecting exactly the opposite current transients defined by a conductance model. This can be used, for example, to find out whether a certain current component is necessary for the generation of action potentials in pacemaker cells (Wilders 2006).

The dynamic clamp method was originally used to simulate the electrical coupling of cardiac muscle cells through gap junctions. Today it is used in neurophysiology to determine how intrinsic conductances and synaptic inputs of neurons affect their behavior. In particular, cardiac- and neurophysiologists use the method to test the mechanisms responsible for the very diverse and characteristic waveforms of action potentials of different cell types (Goaillard and Marder 2006).

Thanks to the increasing speed and performance of computers, it is becoming possible to measure responses to a current injection in near real time, compare them with expected values and correct them. In this way, a closed-loop procedure can be used to enforce a certain behavior of the cell and learn about the underlying conductivity. The dynamic clamp technique thus combines the modeling of cellular behavior with experimental laboratory research.

Some modern amplifiers and their control programs already contain ready-made solutions for this procedure. However, one should not underestimate the effort to master and apply this technique. Preliminary experiments with a model cell are a good way to become familiar with the theory and practice the method.

6.8 In Vivo Patch Clamp

In recent years, research in physiology has increasingly moved toward more complex, physiologically meaningful animal models. Much that was previously possible only in isolated cells is now being studied in organ preparations or even in vivo. This includes applications of the patch clamp technique in living animals, which is now established in many laboratories and has become a standard (albeit very demanding) method for high-resolution cellular electrophysiology with the highest possible biological relevance to the whole system.

Patch clamp recordings in vivo are usually performed in the whole-cell configuration, in addition to the use of patch pipettes for juxtacellular and loose-patch recordings. Alternative or supplementary approaches are recordings with sharp microelectrodes, field potential or *unit* recordings (Sect. 2.2). Patch clamp recordings in vivo are mainly used in neurophysiology. They make it possible to precisely record the subthreshold behavior of individual cells, i.e., changes in membrane potential below the threshold of action potentials, while the neurons are integrated into their natural networks. Of particular interest are postsynaptic potentials (or the corresponding currents in voltage clamp mode). Most commonly, rodents are used, but other species such as insects or fish have also been studied using this method. The first recordings in a living animal were made in the visual cortex of the cat (Pei et al. 1991), while now recordings are also being made from deeper brain regions (e.g., the hippocampus) and from the spinal cord.

The electrophysiological basics are of course the same as for any other preparation, but there are some peculiarities that we would like to touch upon here. Basically, the animal can be studied in three different situations: 1) anesthetized, 2) awake, but restrained, 3) freely moving. The first two versions are by far the most common, while measurements in freely moving animals are the absolute exception. They were developed by Michael Brecht in the lab of Bert Sakmann and provide unique insights into the cellular physiology of animals during behavior (Brecht et al. 2004). But they also require a lot of experience, great experimental skills and considerable frustration tolerance ("considerable" is probably a gross understatement). In this introductory book, we focus on recordings from the restrained animal, where relative movements between the pipette and the brain are largely avoided.

Anesthetized mice or rats are clamped into (commercially available) stereotactic devices that have bars for fixation to the bony skeleton of the ears (external auditory canal) and the upper incisors. Extreme care must be taken during fixation to avoid injury. During prolonged experiments, it is important to protect the animal from cooling down, so it should be kept on commercially available heating pads. Adequate and well-controlled anesthesia is, of course, essential. It must be planned in detail in advance, and in most countries it must be officially approved. By definition, anesthetics induce changes in perception, behavior and neuronal activity, such that there was an early desire to use the technique with non-anesthetized animals. Recordings in awake, head-fixed rodents were also developed in the lab of Bert Sakmann (Margrie et al. 2002). A bar is attached to the head of the animals using dental cement, and the bar is mounted on a special frame for recordings. The animals can run on a conveyor belt or on a moving ball, if necessary even within a virtual reality that simulates a realistic, variable spatial surrounding (Fig. 6.8). Before the actual recordings, the animals are habituated for several days to avoid stress during the experiment.

Access to the target tissue must be created very gently. Too large openings of the skull and the dura lead to strong breath- or pulse-related movements, swelling or dehydration of the brain! The first step is to remove a large enough area of fur to allow for orientation using the cranial sutures. After removing all tissue remnants, a ring can be cemented to the clean and dry bone, forming a small

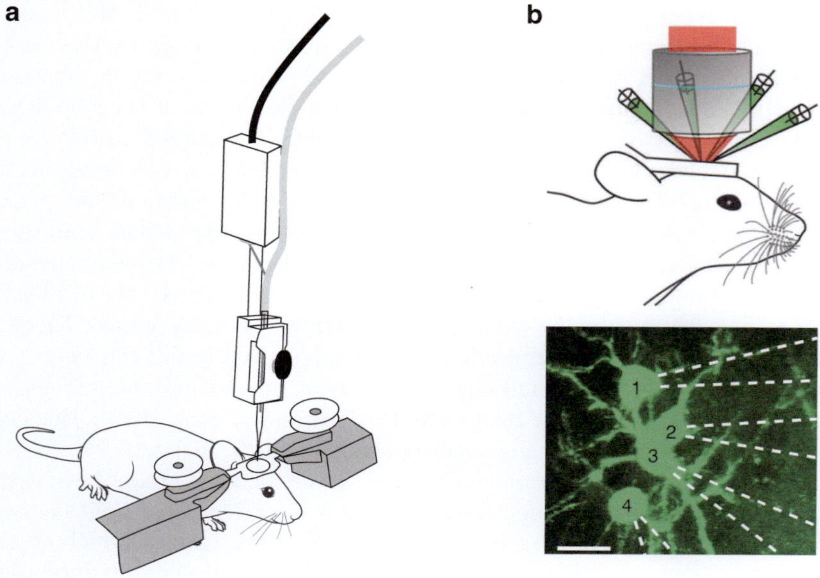

Fig. 6.8 Patch clamp experiments under in vivo conditions. **a** Fixation of the mouse using holders on the head. **b** Pipette positioning under optical control. Below: Multiple recordings of four pyramidal cells in the mouse neocortex filled with a fluorescent dye (Alexa 549) by the pipettes. (**b** From Jouhanneau et al. 2018)

chamber for filling with extracellular solution. There are other approaches without the chamber, but care must be taken to ensure that the surface of the brain does not dry out. Then, under good lighting and control through a binocular or surgical microscope, a small (e.g., 0.5 mm) hole is drilled into the skull at the desired coordinates. The last remaining fine bone lamella can be carefully removed with an injection needle. The underlying hard meninges (dura mater) can also be opened in the minimum size required with a fine needle. Some experimenters leave the dura intact, especially in non-anesthetized animals. A very high internal pressure (e.g., 500 mbar) must then be applied when inserting the pipette, which is drastically reduced after penetrating the dura to avoid tissue damage.

The cell can be accessed "blindly" by slowly advancing the pipette in the axial direction, paying attention to small changes in resistance (test pulse!), then releasing the pressure and gently aspiring the membrane of the suspected cell. Initially, the cell is unknown, but it can be filled with dye in whole-cell recordings and subsequently identified (Sect. 6.11). Another approach is to combine the method with high-resolution optical methods up to 2-photon microscopy. In these cases, a water-immersion objective is lowered into the extracellular solution of the previously created water bath. However, the visual conditions in the native tissue are poor, so previously labeled cells are usually used, which are expressing fluorescent proteins by means of transgenic techniques or viral vectors. This can highlight specific cell types, such as a particular subtype of interneurons. In this case, the pipette should contain a fluorescent dye with a different emission wavelength to be clearly visible. A fluorescent dye (e.g., Alexa 594) leaking from the pipette can also be used to stain the extracellular space, so that the somata of the cells become visible as dark shadows (shadow patching) (Kitamura et al. 2008).

As mentioned above, in vivo patch clamp has become a widely used, but still technically complex application. It requires a lot of preparation and practice and usually results in significantly fewer successful recordings than corresponding measurements in brain slices or cell cultures. Therefore, for many questions, it makes sense not to investigate all mechanistic details in vivo, but to focus on those aspects that require the whole brain or even the behaving animal. There is a trend towards automating the patch clamp technique, where the challenging and time-consuming tasks of identifying suitable cells, approaching the pipette, and seal formation are taken over by a control software (Suk et al. 2019). This could greatly facilitate the dissemination and application of the technique.

For details, we again refer to the literature. Several leading groups have written excellent reviews on the method (e.g., Noguchi et al. 2021). This applies not only to rodents but also to insects, fish, and other organisms, each of which requires special conditions for fixing and care. We would also like to point out the use of so-called whole-brain preparations of small animals, which can be easily kept alive for the duration of an experiment. Even in mammals, it has been possible to keep a removed brain functionally intact by perfusing the arteries, although this technique has so far been largely limited to guinea pigs (de Curtis et al. 2016).

6.9 Multiple Recordings

Patch clamp recordings from multiple cells simultaneously allow the parallel registration of cell-specific signals under identical stimulation, activity, or pharmacological conditions. A special feature is that pre- and postsynaptic mechanisms can be studied at synaptic connections between cells of the same or different cell types, and information about the type, strength, and frequency of synapses can be collected (Fig. 6.9). However, even finding and stably recording from just two connected cells in a patch clamp experiment is often a major challenge, especially when the probability of finding connected pairs is low. For this reason, multiple recordings are used to increase efficiency, to find connections more easily, and to speed up the connectivity mapping of local networks. The continuous development of small manipulators and pre-amplifiers now allows parallel recordings from large numbers of cells. Here, even up to ten simultaneous whole-cell measurements are no longer a rarity (Espinoza et al. 2018; Peng et al. 2019).

The technical requirements for multiple recordings are, of course, first a sufficient number of pipette holders and amplifiers (multichannel amplifiers are advantageous, but not absolutely necessary). However, additional measures are needed to reliably patch multiple cells without losing established recording configurations. Here are some useful tips: The microscope should be freely movable with respect

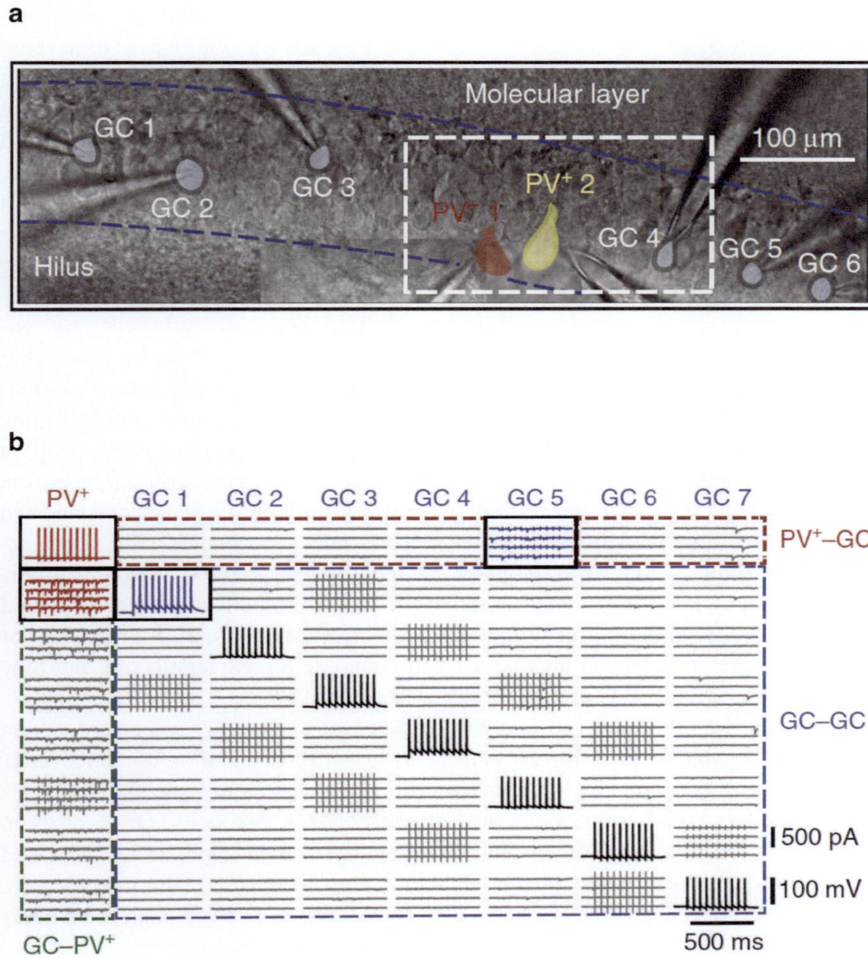

Fig. 6.9 Multiple recordings. **a** Seven simultaneous whole-cell recordings of granule cells and two inhibitory interneurons in the dentate gyrus of a mouse (brain slice). **b** Connection matrix between seven granule cells and a parvalbumin-expressing interneuron. The diagonal line shows the intrinsic properties of the presynaptic neuron, the other parts show the postsynaptic currents of the postsynaptic neurons. (Modified after Espinoza et al. 2018)

to all other components (X-Y table, Sect. 4.1.2), in order to approach cells outside the current field of view and to place additional pipettes. In addition, manipulators driven by electric motors should be sufficiently mechanically decoupled to avoid the transmission of vibrations to pipettes in the tissue. Care (or rather a delicate touch) is also required when inserting the pipette and approaching the cell. In practice, slight movements of the already inserted pipettes are unavoidable. In the whole-cell configuration, however, the access usually remains stable despite such small movements. Many common patch clamp rigs are designed for multiple independent manipulators, so all you need are good specimen and the willingness to develop your own method and sequence for making multiple recordings through practice.

Two typical mishaps are almost inevitable: Sometimes, you use the wrong manipulator control and push a pipette that is already in the tissue into the bottom of the chamber instead of bringing a new one under the objective. Another time, you are suddenly met with a broken pipette that crashed into a previously placed pipette in the tissue. As a remedy, we recommend moving the objective up as far as possible, while accepting a longer “approach” of the pipette to the cell in order to avoid such collisions.

If two or more cells are successfully brought into the whole-cell mode, the synaptic or electrical connections can be tested. This is usually done by generating action potentials in one of the cells in current clamp mode. In the case of a monosynaptic connection, the postsynaptic signals of the other cell(s) should follow with a short latency. The magnitude, time course and temporal offset as well as its variability depend strongly on the type of synapses involved. Important factors include the type of transmitter system, the release probability of vesicles and the receptors and ion channels involved on the pre- and postsynaptic sides. In practice, the resulting postsynaptic signals are recorded in whole-cell mode in current or voltage clamp. For some research questions, the perforated-patch method (Sect. 6.2) is a good option for both the pre- and postsynaptic cell. Juxtacellular or cell-attached recordings can also be used to stimulate a presynaptic cell, but they do not provide the stability and control of a whole-cell access.

When action potentials in a neuron generate temporally precisely correlated synaptic signals in a downstream cell, this is referred to as unitary events, to describe the smallest unit of signal transmission. It should be noted that typically several synaptic connections are formed between two cells. Differences in the activity and efficiency of the individual synapses then generate an increased variability in the postsynaptic responses, which may need to be taken into account. Often, however, the synaptic connection is considered as a whole, and the signals are evaluated as the sum of these different individual contributions.

Paired recordings of two connected cells offer good experimental conditions for the targeted manipulation of synaptic transmission, since modulating substances as well as neurotransmitters can be introduced into a single cell in defined concentrations. Thus, intracellular processes such as neurotransmitter metabolism, vesicle loading and their release in interaction with calcium can be studied. Especially the direct patching of a large presynapse together with the postsynaptic cell allows a very effective control of the intracellular milieu and the electrical activity (shape and sequence of action potentials) of the connection (e.g. the Calyx of Held in the brainstem). Even smaller presynaptic structures in the cortex or hippocampus, such as mossy fiber boutons, can nowadays be reliably stimulated in the cell-attached or even in the whole-cell mode with the help of some tricks and additional components, in order to directly study a cortical presynapse together with the postsynaptic cell (Vandael et al. 2021).

6.10 Automated Patch Clamp

As with all electrophysiological methods, the patch clamp technique requires special skills and care in handling the tissue. This is what makes this field of research so attractive to many of us, but it also leads to relatively low sample sizes ("n numbers"). In a series of recordings using pharmacology, it can take a long time to test each of perhaps 5 concentrations of a substance on 15 or 20 cells. This quickly becomes a limiting factor when examining multiple substances. Therefore, there was an early desire to develop automated procedures with higher throughput (Fig. 6.10) (Dunlop et al. 2008).

In Sect. 6.8 we have already discussed software solutions for in vivo recordings that control the automatic approach of the pipette onto the cell and the creation of a seal. For isolated cells in suspension, the even further automated flip-tip technique has been developed, in which the cells are filled into a patch pipette from the back together with the extracellular solution and sink down until a cell finally blocks the opening (Fejtl et al. 2007). The pipette is then placed into a reservoir (tip socket) with intracellular solution, which is connected to a patch clamp pre-amplifier and a suction system. Negative pressure is then applied until a seal is formed and finally ruptured. These patch clamp robots allow parallel recordings with multiple pipettes, so that several hundred cells can be measured per day. Other systems pull the cells to a small hole in the bottom of a chamber, under which a patch pipette is placed that automatically creates the seal. However, the throughput of such techniques based on individual glass pipettes is always limited.

For high-throughput screening, another method has been established in which isolated cells in a suspension are aspired onto a perforated plate by negative pressure. Originally designed for single cells, these systems were quickly further developed into multi-well planar array plates that allow parallel measurements of 16, 64 or more cells. This allows efficient pharmacological screening, although the quality of the data is unlikely to match that of single pipette recordings. The materials of the well plates and the perfusion systems vary from manufacturer to manufacturer. Larger systems can record several thousand cells per day. For this purpose, mostly electrophysiologically (largely) neutral cells (HEK, CHO etc.) are used, into which defined ion channels have been introduced by transfection.

6.11 Staining After the Recording

After a successful recording, one often wants to examine the recorded cell more closely histologically. Typical questions are the classification of the cell type, the size and degree of branching of the cells, localization within the tissue or synaptic connectivity (especially when recording from multiple cells). This is usually achieved by introducing a dye via the patch pipette. In neuroscience, biotin derivatives are mostly used (biocytin or neurobiotin), to which the protein avidin (or

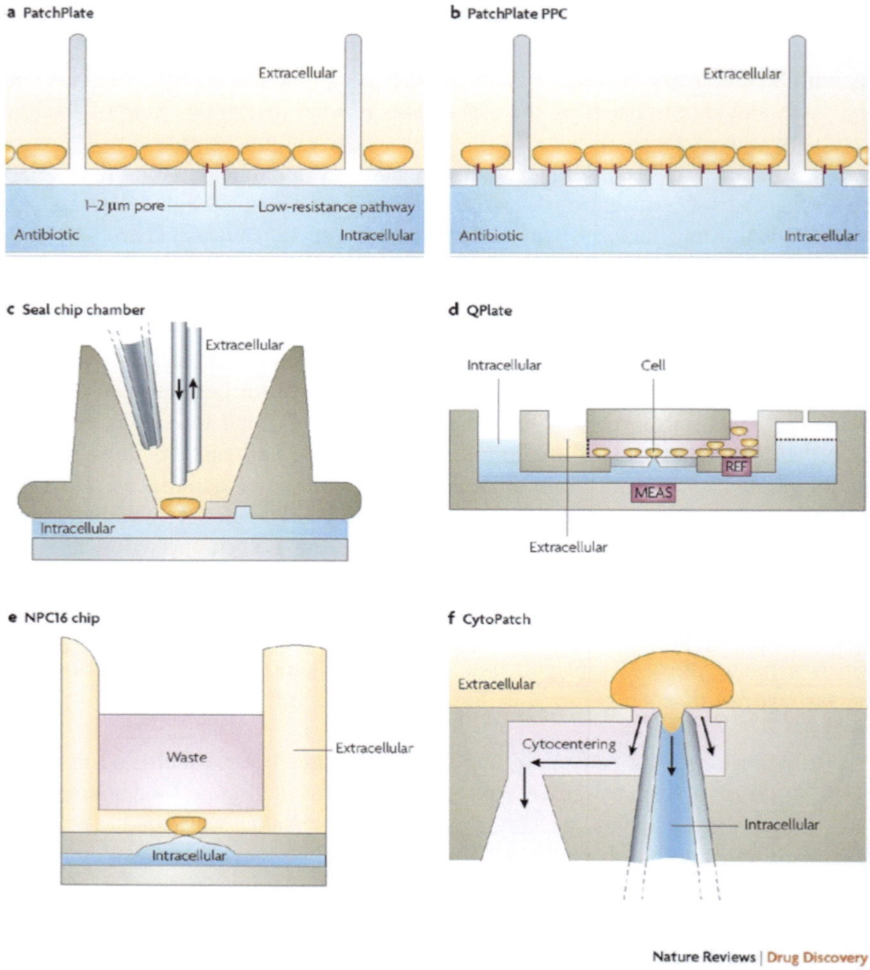

Fig. 6.10 Automated patch clamp. Cells are placed in various configurations at openings that provide access either by perforated patch technology (Sect. 6.2) (**a, b**) or conventional whole-cell recording (**c–f**). Details on the different technologies in the original article. (Dunlop et al. 2008)

streptavidin) subsequently binds to. Visualization can be achieved by labeling with fluorescent dyes. For stable and very high-contrast stainings, peroxidase can be used, which, together with hydrogen peroxide and coloring metals, allows a dark brown or black staining of the cell (Fig. 6.11) (Marx et al. 2012).

The result is a stained cell in the tissue compound, which is contrasted by a background stain (Nissl, DAPI) to provide a clear orientation of nuclei, cell layers, etc. The combination with other immunostaining techniques is also possible – protocols are published or provided by the manufacturers.

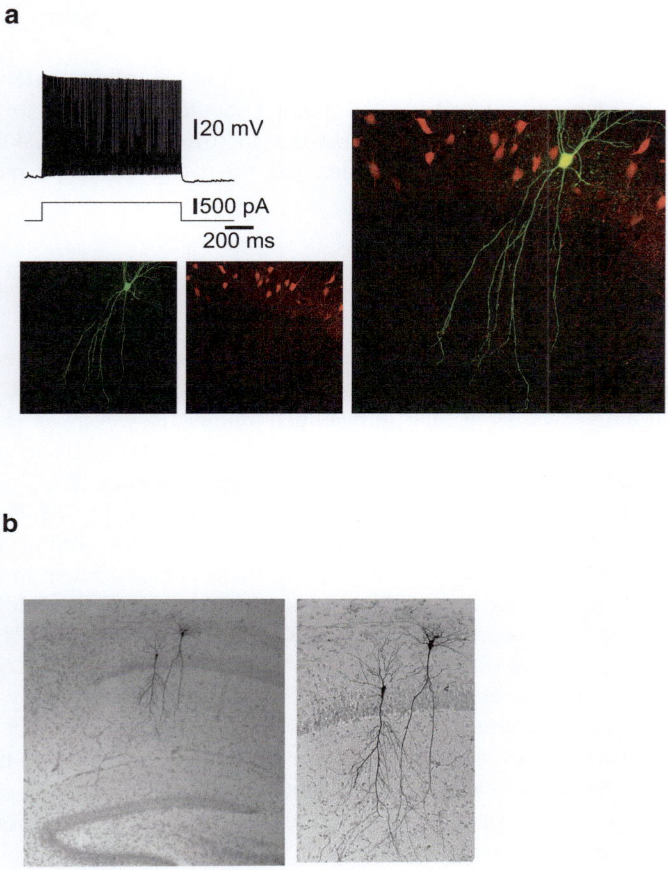

Fig. 6.11 Immunostaining after a patch clamp recording. **a** Intrinsic properties (whole-cell, current clamp) and morphology of a parvalbumin-positive interneuron of the CA1 region in the murine hippocampus. The cell was filled with biocytin during the recording and then imaged by fluorescence microscopy. Red labeled cells show the population of parvalbumin-expressing interneurons (transgenic mouse line, expression of Rosa-tdTomato). **b** Conversion of two biocytin-filled cells into a (non-fluorescent) staining by coupling to peroxidase and reaction with diaminobenzidine (DAB)

A prerequisite for the staining and morphological analysis of the examined cells is, however, that the pipette is carefully removed from the cell so that the membrane can close again. If the movements are too rough or too fast, either the soma and the structures near the soma will be destroyed, or the cell will be pulled through the tissue together with the pipette, injuring and displacing the cell. After detaching the pipette, it is advisable to wait a while before fixing the tissue, especially after short recordings (the required time depends on the size and complexity of the cell and ranges from a few minutes to about 1 h). This will allow the dye to distribute to the periphery of the cell, hopefully resulting in complete and uniform staining.

For fixation, a 1–4% paraformaldehyde solution in phosphate buffer is usually used. The duration and intensity of the fixation should be adjusted to the subsequent procedures. Some antigens are lost due to (over-)fixation, making certain subsequent immunostainings impossible, and the tissue becomes more cross-linked and less permeable to antibodies.

6.12 Single-Cell RT-PCR and Patch-Seq

In the whole-cell configuration, the pipette has direct access to the interior of the cell. We have emphasized so far that this allows the pipette solution to enter the cell and virtually "wash out" the intracellular milieu. However, as early as the 1990s, scientists recognized that this also provides an opportunity to collect and analyze valuable molecules from the cell under investigation. This is especially true for mRNA, which reflects the transcriptional activity of the cell. By carefully aspirating the cytoplasm into the pipette at the end of a whole-cell measurement, one can in principle determine the mRNA sequences contained therein and obtain information about the gene expression of the examined cell in addition to the electrophysiological properties (Eberwine et al. 1992; Audinat et al. 1996). The technique can be used to distinguish between different neurons or to correlate the functional data with the expression of certain proteins (**caution:** only mRNA is detected, conclusions about the existing proteins are therefore indirect and especially quantitatively limited). Typical research questions ask whether the expression of critical proteins is altered in a disease model, whether a cell expresses a specific subunit of a receptor or ion channel, or whether a "marker" for a certain cell type is present (e.g., parvalbumin, which is characteristic for a special class of inhibitory interneurons). A more recent, extended application is known as patch-seq: the goal is to determine the entire RNA of the aspired cytoplasm, i.e. the entire transcriptome of the cell, which, together with other functional and structural data, allows a very precise differentiation of cell types.

The principle is to generate the much more stable DNA from the collected mRNA by reverse transcription (RT) and to amplify the sequences of interest from this by polymerase chain reaction (PCR), which are then characterized with conventional molecular biological techniques and assigned to individual RNA species. The experimental procedure consists of 1) cytoplasm collection, 2) reverse transcription, 3) amplification of present sequences by PCR, and 4) analysis of the DNA species obtained (Fig. 6.12).

Due to the small amounts and instability of RNA, it is important to keep all parts that might come into contact with the RNA free of RNA-degrading enzymes (RNases), which are virtually ubiquitous. Meticulous work is the key to this method. Even with the greatest precautions, the experiment is usually kept as short as possible to minimize degradation of the RNA. The success of the method

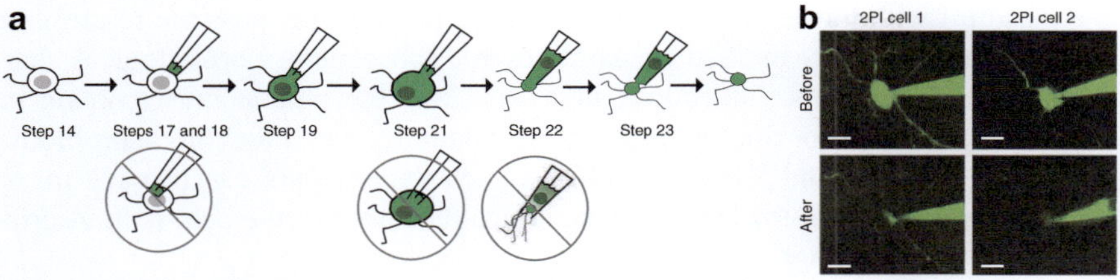

Fig. 6.12 Procedures for single-cell RT-PCR. **a** Schematic of the lateral approach of the pipette and the harvesting of the cytoplasm during and after whole-cell recording. The circled images show possible errors: non-central pipette position, too strong indentation of the cell and too strong aspiration, which makes subsequent morphological analysis impossible. **b** Original photos of a recording with aspiration of the cytoplasm (fluorescence images, filling with Alexa-488). (From Cadwell et al. 2017)

depends on the strict RNase-free nature of all critical solutions and surfaces. This includes in particular the patch pipette, the silver wire, the intracellular solution, the pipette tips, and anything that may come into contact with the intracellular solution beforehand (glasses, etc.). The intracellular solution is autoclaved after preparation, pressed through a filter (RNase-free syringe!), aliquoted and stored at −20 °C. Some authors recommend protecting the pipette tip from contamination by dipping it briefly into a protein-repellent solution (organopolysiloxane; so-called Sigmacote). The pipettes are filled with the smallest possible amount of intracellular solution (less than 10 µl may be sufficient!) using a precision pipette to keep the dilution of the RNA as low as possible. When approaching the cell, it is important to maintain a constant positive pressure to prevent contamination of the pipette interior. The pipette opening should be as wide as possible to facilitate the aspiration of the cytoplasm.

After the recording is completed and the cytoplasm has been aspirated, the pipette is carefully withdrawn and the contents are emptied into a prepared reaction tube. This is done by breaking off the tip of the pipette at the bottom of the tube and applying positive pressure to the inside of the pipette from the back using a syringe and tubing. Buffer, enzymes, and substrates for reverse transcription are already in the tube, so that the RNA is transcribed into cDNA, which is much more stable. The sequences of interest can then be amplified by PCR. You can use primers for very specific sequences ("does my cell express the subunits X, Y or Z of a certain ion channel?"), or you can work with "random primers", so that you amplify the entire RNA and then analyze it to the desired depth using agarose gels, arrays or sequencing devices (*patch-seq*).

The analysis can also be performed quantitatively as so-called qPCR, although the "harvesting" of the RNA from the cytoplasm naturally limits the accuracy of any subsequent analysis. Errors are greatly amplified by the exponentially increasing amount of cDNA in the PCR cycles. There are protocols to improve the accuracy of qPCR using linear methods of amplification – please refer to the relevant literature.

By careful withdrawal and detachment of the pipette, it is possible to preserve the neurites of neurons in slice preparations for subsequent morphological characterization of the cells. Of course, in these cases the pipette must contain an appropriate indicator for fluorescence or fixed staining (e.g., biocytin, neurobiotin) (Sect. 6.11) (Kamen and Karadottir 2021). Such experiments can provide maximum information about the electrophysiological behavior of the cell, its transcriptome, and its morphology.

References

Anson BD, Roberts WM (2002) Loose-patch technique. In: Walz W, Boulton AA, Baker GB (eds) Patch-clamp analysis: Advanced techniques. Humana Press, Totowa, NJ, pp 265–286

Aquila M, Benedusi M, Fasoli A, Rispoli G (2015) Characterization of zebrafish green cone photoresponse recorded with pressure-polished patch pipettes, yielding efficient intracellular dialysis. PLoS ONE 10:e0141727

Audinat E, Lambolez B, Rossier J (1996) Functional and molecular analysis of glutamate-gated channels by patch-clamp and rt-pcr at the single cell level. Neurochem Int 28:119–136

Brecht M, Schneider M, Sakmann B, Margrie TW (2004) Whisker movements evoked by stimulation of single pyramidal cells in rat motor cortex. Nature 427:704–710

Cadwell CR, Scala F, Li S et al (2017) Multimodal profiling of single-cell morphology, electrophysiology, and gene expression using patch-seq. Nat Protoc 12:2531–2553

de Curtis M, Librizzi L, Uva L (2016) The in vitro isolated whole guinea pig brain as a model to study epileptiform activity patterns. J Neurosci Methods 260:83–90

Daniel J, Polder HR, Lessmann V, Brigadski T (2013) Single-cell juxtacellular transfection and recording technique. Pflugers Arch 465:1637–1649

Davie JT, Kole MH, Letzkus JJ, Rancz EA, Spruston N, Stuart GJ, Hausser M (2006) Dendritic patch-clamp recording. Nat Protoc 1:1235–1247

Dunlop J, Bowlby M, Peri R, Vasilyev D, Arias R (2008) High-throughput electrophysiology: An emerging paradigm for ion-channel screening and physiology. Nat Rev Drug Discov 7:358–368

Eberwine J, Yeh H, Miyashiro K et al (1992) Analysis of gene expression in single live neurons. Proc Natl Acad Sci U S A 89:3010–3014

Espinoza C, Guzman SJ, Zhang X, Jonas P (2018) Parvalbumin(+) interneurons obey unique connectivity rules and establish a powerful lateral-inhibition microcircuit in dentate gyrus. Nat Commun 9:4605

Fejtl M, Czubayko U, Hummer A, Krauter T, Lepple-Wienhues A (2007) Flip-the-tip: Automated patch clamping based on glass electrodes. Methods Mol Biol 403:71–85

Gillis KD (2000) Admittance-based measurement of membrane capacitance using the epc-9 patch-clamp amplifier. Pflugers Arch 439:655–664

Goaillard JM, Marder E (2006) Dynamic clamp analyses of cardiac, endocrine, and neural function. Physiology (Bethesda) 21:197–207

Gurkiewicz M, Korngreen A (2006) Recording, analysis, and function of dendritic voltage-gated channels. Pflugers Arch 453:283–292

Horn R, Marty A (1988) Muscarinic activation of ionic currents measured by a new whole-cell recording method. J Gen Physiol 92:145–159

Hu W, Shu Y (2012) Axonal bleb recording. Neurosci Bull 28:342–350

Jouhanneau JS, Kremkow J, Poulet JFA (2018) Single synaptic inputs drive high-precision action potentials in parvalbumin expressing gaba-ergic cortical neurons in vivo. Nat Commun 9:1540

Kamen Y, Karadottir RT (2021) Combining whole-cell patch clamp and dye loading in acute brain slices with bulk rna sequencing in embryonic to aged mice. STAR Protoc 2:100439

Kim S, Guzman SJ, Hu H, Jonas P (2012) Active dendrites support efficient initiation of dendritic spikes in hippocampal ca3 pyramidal neurons. Nat Neurosci 15:600–606

Kitamura K, Judkewitz B, Kano M, Denk W, Hausser M (2008) Targeted patch-clamp recordings and single-cell electroporation of unlabeled neurons in vivo. Nat Methods 5:61–67

Lindau M (2012) High resolution electrophysiological techniques for the study of calcium-activated exocytosis. Biochim Biophys Acta 1820:1234–1242

Margrie TW, Brecht M, Sakmann B (2002) In vivo, low-resistance, whole-cell recordings from neurons in the anaesthetized and awake mammalian brain. Pflugers Arch 444:491–498

Marrero HG, Lemos JR (2007) Loose-patch-clamp method. In: Walz W (ed) Patch-clamp analysis: Advanced techniques. Humana Press, Totowa, NJ, pp 325–352

Martina M, Schultz JH, Ehmke H, Monyer H, Jonas P (1998) Functional and molecular differences between voltage-gated k+ channels of fast-spiking interneurons and pyramidal neurons of rat hippocampus. J Neurosci 18:8111–8125

Marx M, Gunter RH, Hucko W, Radnikow G, Feldmeyer D (2012) Improved biocytin labeling and neuronal 3d reconstruction. Nat Protoc 7:394–407

Neher E (2006) A comparison between exocytic control mechanisms in adrenal chromaffin cells and a glutamatergic synapse. Pflugers Arch 453:261–268

Neher E, Marty A (1982) Discrete changes of cell membrane capacitance observed under conditions of enhanced secretion in bovine adrenal chromaffin cells. Proc Natl Acad Sci U S A 79:6712–6716

Noguchi, A., Ikegaya, Y., Matsumoto, N. (2021) In vivo whole-cell patch-clamp methods: Recent technical progress and future perspectives. Sensors (Basel) 21.

Pei X, Volgushev M, Vidyasagar TR, Creutzfeldt OD (1991) Whole-cell recording and conductance measurements in cat visual cortex in-vivo. NeuroReport 2:485–488

Peng Y, Mittermaier FX, Planert H, Schneider UC, Alle H, Geiger JRP (2019) High-throughput microcircuit analysis of individual human brains through next-generation multineuron patch-clamp. Elife 8.

Pinault D (1996) A novel single-cell staining procedure performed in vivo under electrophysiological control: Morpho-functional features of juxtacellularly labeled thalamic cells and other central neurons with biocytin or neurobiotin. J Neurosci Methods 65:113–136

Rhee JS, Ebihara S, Akaike N (1994) Gramicidin perforated patch-clamp technique reveals glycine-gated outward chloride current in dissociated nucleus solitarii neurons of the rat. J Neurophysiol 72:1103–1108

Stuart GJ, Dodt HU, Sakmann B (1993) Patch-clamp recordings from the soma and dendrites of neurons in brain slices using infrared video microscopy. Pflugers Arch 423:511–518

Suk HJ, Boyden ES, van Welie I (2019) Advances in the automation of whole-cell patch clamp technology. J Neurosci Methods 326:108357

Tang JM, Wang J, Quandt FN, Eisenberg RS (1990) Perfusing pipettes. Pflugers Arch 416:347–350

Vandael D, Okamoto Y, Borges-Merjane C, Vargas-Barroso V, Suter BA, Jonas P (2021) Subcellular patch-clamp techniques for single-bouton stimulation and simultaneous pre- and postsynaptic recording at cortical synapses. Nat Protoc 16:2947–2967

Wilders R (2006) Dynamic clamp: A powerful tool in cardiac electrophysiology. J Physiol 576:349–359

Zhu WJ, Vicini S (1997) Neurosteroid prolongs gabaa channel deactivation by altering kinetics of desensitized states. J Neurosci 17:4022–4031

Data Handling and Documentation 7

Performing patch clamp experiments is initially a very "hands-on" task. When everything finally "works", there is often a temptation to perform many recordings in a row, possibly with slight variations in the recording conditions. Quite a few (including us) have found themselves after weeks of intensive work in front of mountains of quite heterogeneous raw data to analyze. This chapter is intended to help keep an eye on the acquired data from the very beginning, so that you get consistent data sets that are easy to analyze, communicate and share with others.

We will start quite pragmatically with the experiment itself, then give advice on data processing and storage, and finally discuss the trend toward transparency and free availability of data, which is increasingly determining current standards of good scientific practice.

7.1 Documentation of Laboratory Work

At the beginning, there is the simple insight that you can only scientifically evaluate and communicate what you have documented beforehand. This may sound trivial, but we all have made some mistakes in this regard and therefore want to address this point explicitly.

Good scientific practice includes the documentation of all laboratory work in a laboratory notebook (or "laboratory journal"). These notebooks, as well as the storage of all raw data must be secured for 10 years. Carefully stored and daily updated bound paper notebooks are fine, but digital solutions are increasingly being used. Such electronic lab notebooks (ELN) are available from both commercial vendors and open-source projects. Choosing the right product is not easy, especially since the field is rapidly evolving. Many research institutions are adopting ELNs as a standard and provide them as part of the research infrastructure. Other research institutions (still) strictly require physical lab books. In this case, it

F. C. Roth et al., *Mastering the Patch Clamp Technique: A Practical Guide*, https://doi.org/10.1007/978-3-662-70745-6_7

is worth considering whether an ELN should be created in parallel, since the electronic storage of metadata associated with the raw data and other information can greatly facilitate work.

All lab books must allow for time-stamped, unalterable documentation of the work. Paper versions should therefore contain page numbers (and all entries should be dated), electronic lab books should store the entries in such a way that they can be traced. This strict chronology must never be broken, i.e., changes are not made by overwriting the old content, but by dated additions that are clearly separated from the previous ones. Good practice also includes regular acknowledgement and signature by the research group leader. Such routines ensure clarity about the current state of the project and strengthens the collaborative work on the project. They also provide clear information about what needs to be documented. In addition, they automatically create security and transparency regarding all collected data, without raising a climate of surveillance or even suspicion, which has no place in a good research group.

What should be in the lab book? In general, all relevant work in the lab must be documented. This includes not only the patch clamp experiment itself (see below), but also the preparation of new intra- or extracellular solutions, stock solutions of salts or drugs, the preparation and maintenance of cultured cells, the preparation of an animal by injection of viruses, the implantation of electrodes or behavioral training—in short: everything that is or could be relevant to the obtained results. Useful hints can be found in good laboratory practice guidelines or method books.

Many methods become routine during the course of a project and only need to be noted briefly together with experimental details under the relevant date. For example, the preparation of a new batch of extracellular solution will be documented, but the recipe will not be written down every single time. However, two things should be noted: 1) The recipe must be stored somewhere in a clear, permanent and easily retrievable manner, preferably (also) in electronic form. 2) If something has changed in the routine (e.g. a change in the calcium or magnesium concentration), it must of course be explicitly documented. It must be completely clear up to which day recordings were carried out with the old solution and from which day the new one has been used (whereby "old" and "new" would be the worst possible names for the respective solutions, which will become clear at the latest after a further renewal).

It is important that all batches and versions are documented in the lab book. When did I make a new stock solution of a regularly used drug, when did I order and use a substance for the first time? Which batch of cultured cells was used in the experiment? Which batch of a virus did I inject stereotactically into the animals? Which program version was used for a recording or analysis? Such information is very helpful at later stages, especially when inconsistencies appear in the data.

Finally, the documentation of animal experiments is of course subject to particularly strict rules. Since the legal requirements for such experiments vary from country to country, we do not describe the procedure in detail here. However, it should be noted that the lab book must clearly identify each animal used. Electronic lab books allow for the integration of photos, so that, for example, the cage card (or photocopies for the paper version) can be included for verification.

7.1.1 Documentation of a Patch Clamp Experiment

The data from a patch clamp experiment consists of three main components:

1. Raw current and voltage data,
2. other associated data (e.g., live imaging of calcium signals or video recordings of the behaving animal),
3. Metadata containing all other information (e.g. date, type of experiment, batch of cells, identity of the animal used, peculiarities of the recording protocol, device settings). If important co- or assistance work by other persons was involved, this should also be documented.

All of this data must be kept for at least ten years, in a way that it can be easily retrieved and all related information (especially raw and metadata) remains linked. Unfortunately, there is no one-size-fits-all solution but there are some rules that can make life much easier.

Raw data of currents and voltages: The digitized current and voltage values from the patch clamp experiment are usually stored in a proprietary file format of the respective recording program. In general, we recommend storing more data rather than less, i.e. also failed experiments, routine recordings of passive and active membrane properties (Sect. 2.1), sufficiently long baseline recordings before an intervention, the wash-out phase of substances, etc. Storage space in the relevant order of magnitude is not expensive, so it is always worthwhile to record as much as possible. If unexpected questions arise during the analysis, you will certainly be grateful for this completeness.

▶ Many experiments fail, i.e. no usable recording is obtained, or the cell does not show the functions you want to study. Often such experiments are simply ignored. But sometimes the question arises, in how many cases an experiment was successful at all. How many of the recorded neurons showed postsynaptic potentials when afferent fibers were stimulated? How many cells were recorded without a certain current component being detectable (think of the study of a loss-of-function mutant!). In such cases, it is very good to have documented the "failed attempts" and to be able to state the success rate of the experiment.

Associated data: Heterogeneous, multimodal data sets from individual experiments are common in electrophysiology. In addition to the patch clamp data, other signals and parameters are often recorded, e.g. extracellular potentials, force measurements of muscle cells, fluorescence images of calcium-sensitive dyes. Some of these data can simply be recorded on additional channels of the same analog-to-digital converter, and processed by the recording software used for the patch clamp experiment. This greatly facilitates evaluation based on a common timeline - the (correctly named and calibrated) data are immediately available in parallel to the patch clamp data and can be correlated with them. Example: How did the membrane currents of a neuron (patch clamp) behave during a network oscillation of the relevant brain region (extracellular field potential)?

Things become more complex when data are collected independently and stored in different programs, file formats, or with different temporal resolution (different sampling rate). This is particularly common when electrophysiology is combined with imaging techniques. However, completely different data can also be added, e.g. results from behavioral experiments or histological findings. For real-time measurements (live imaging, time-critical behavioral tests, etc.), the problem of precisely synchronizing the separately stored data sets in the analysis is often encountered. Although there are software solutions for integrating heterogeneous data and precisely aligning data sets, individual or even manual solutions are often used. It is important to clarify beforehand how precise the temporal assignment should be—do I want to link a cellular reaction to an electrical phenomenon with millisecond precision, or is it enough to know that substance X was slowly washed into the bath after 10 min of stable baseline recording and washed out again after another 20 min? Depending on the requirements, there are different solutions.

In the “softer” cases (slow substance application), manually entered markers in the recording program or even notes of the time points in the lab book (not recommended as the sole documentation) can be used. In the typical control programs for patch clamp experiments, the time of a manipulation can be recorded on a separate “marker” channel. Many independent components, such as the stimulation or application devices, can generate 3–5 V square pulses (so-called TTL pulses), which are fed into the recording software of the patch clamp experiment on a separate channel. In optical measurements, a flash of light can perform similar functions, as long as it does not interfere with the experiment (use infrared if possible). Such signals can help to synchronize different modalities retrospectively. These solutions are often complex and inaccurate, but not always avoidable. However, the technical development is moving in the direction of merging heterogeneous data as precisely and standardized as possible—corresponding software or scripts have been published (e.g. Spacek et al. 2008; Sellers et al. 2021; Szell et al. 2020; Klumpp et al. 2025) and are usually freely available on the websites of research institutions or in repositories (pypi.org, open-neuroscience.com, github, MATLAB Central File Exchange).

The type and amount of associated data vary greatly depending on the experimental approach, resulting in very different but related files. It is important that the assignment between the different data sets is clear, so that the patch clamp data of a cell can be unambiguously assigned to the corresponding data from camera recordings (live imaging) or to later added histological images. Meaningful file names and a clear directory structure when saving will help to achieve this (see below).

Metadata: Modern software solutions for controlling patch clamp recordings already store information beyond the actual recorded data points, i.e. so-called metadata. Usually, parameters such as the calibration of the individual channels, amplification factors, filter settings in the program, and sampling rate, etc. are automatically recorded. However, this should be double checked! Also, there are many other factors that are not directly included in the recorded files and must be recorded either in the program's comment fields or separately in the lab book. Here are some suggestions for your own lists:

- Date
- Possibly time of day (circadian rhythm, daily temperature variations!)
- Type of experiment, series of recordings
- Name(s) and path of the respective saved file(s) containing the original data
- If applicable, identity of the animal, approval number of the animal experiment
- Preparation (cell culture batch, organ slice level etc.)
- Type of extra- and intracellular recording solutions used and possible additives
- Incubation protocol of the preparation and storage- and recording chamber used
- Perfusion rate and temperature of the preparation
- For complex preparations: location of the cell(s) and electrodes—it is recommended to use markings in a (pre-made) sketch
- Assignment of channels in the program
- All device settings that are not automatically recorded by the program:
  - Settings of external filters
  - Settings of external pre-amplifiers
  - Values of capacitance and bridge compensation (and their changes during the course of the experiment!)
  - Set parameters (duration, strength) of electrical or optical stimulation devices
- External factors, e.g. room temperature (very important if this changes during the day or seasonally within longer series)
- Name and storage location of histological preparations resulting from the experiment
- Any manipulations during the experiment (e.g. application of substances)

It is important to complete your checklist consistently for each experiment. It is best to create a simple form (electronically or on paper) so that you don't forget anything in the heat of the experiment (Fig. 7.1).

Date: ____________________ Project: ____________________

Name: ____________________ Experiment: ____________________

File path: ____________________ File prefix: ____________________

Permit no. : ____________________ Internal sol.: ____________________

Species: ____________________ External sol.: ____________________

Animal ID: ____________________ Preparation: ____________________

Date of birth: ____________________ Incubation: ____________________

Perfusion speed: ____________________ Temp. in rec. chamber: ____________________

Channel/ADC-port assignments: ____________________ Sampling rate: ____________________

Gain: ____________________

Filter:

File no.	Channel	Mode	RMP	V(c)/V(m)	I(h)/I(c)	Cap	Rs
	Notes:						
	Notes:						
	Notes:						
	Notes:						

Offset change after exp.:

Channel no.:		
Deviation (mV):		

Fig. 7.1 Fill out form for a patch clamp experiment

7.2 Where to Store What?

Related data belongs together! It is important to keep track of the raw data and its corresponding metadata (both the patch clamp recordings and other types of data). There are first approaches to achieve this with the help of electronic lab books, which can store smaller data files and at the same time refer to the respective storage location of the large data files. However, there is no standard solution yet, so we limit ourselves to pointing out that experimental and metadata should be stored in a way that preserves the relationship between all relevant files. This includes cleverly chosen filenames, and possibly lists of each set of related data. We strongly advise against loose paper notes. Lab books are better, digital forms or tables provide an overview, the integration of all information in an electronic lab book is best.

For the storage of the actual electrophysiological data there are some rules that follow from the above:

- Use a clear, uniform naming convention, e.g. your own name abbreviation, date (for automatic sorting by date of the recording use the order year, month, day), number of the cell or the recording ("NN_YYYYMMDD_No.File-ext")
- If you are going to do a lot of processing during the recording (e.g. by filtering) you should also save the unedited raw data. If complications or new questions arise, you can then perform a completely new (digital) filtering and analysis.
- Usually, you will first save the data on a local "recording computer", which also runs the control and recording software for the patch clamp experiment. Later (preferably on the same day!), data is stored on a separate server, which is either part of the lab or a higher-level infrastructure (e.g. university data center). It is important that you are aware of the location, storage duration and security (backup policy!) of this central data storage.
- Once the transfer to the server has been completed and verified (!) you can empty the local lab computer. Otherwise, it will inevitably fill up and become very confusing, especially if it is used by several people.
- At the latest when you start to store data centrally beyond the lab computer, you should use a clear, easily understandable structure of file and folder names. Data from the same series of experiments belong together, and the corresponding directories should have intuitive names.

▶ Over time, large amounts of data accumulate, not all of which need to be processed immediately. The routinely used server will then either reach its limits, or high fees will be charged for storage space. If different storage options are available, it is recommended to separate pure storage from data that is currently being analyzed. Mass storage with high capacity, but lower or slower availability of data, offers a cheap solution for long-term storage (consider 10 years of documentation obligation). The currently processed data is then additionally stored on an "analysis server" with fast data availability. Sometimes local or supra-national data centers also offer servers with very high computing capacity that can be used as "number crunchers" for particularly complex analyses.

7.3 Open Data and Transparency

In recent years, there has been a growing demand claiming that scientific data—at least those from public institutions—should be freely accessible to the public. There are several reasons for this. "Open Science"

- contributes to a more complete picture of the state of knowledge in a field,
- counteracts the widespread ignorance towards "negative results" and reduces the corresponding distortion of the data situation,
- allows new types of analysis by third parties that the original experimenters did not consider,
- allows data from many different labs to be combined into large, very powerful data sets that enable entirely new conclusions to be drawn,
- protects the scientific community from the scientific misconduct of individuals,
- protects experimenters from accusations of scientific fraud.

All this speaks in favor of making one's data generally available, at the latest together with the corresponding publication. Before that, one might not want to draw the competitor's attention to one's own results. This problem is mitigated by open-access repositories such as BioRxiv or MedRxiv, the online archives of the Cold Spring Harbor Laboratory (https://www.biorxiv.org/ or https://www.medrxiv.org/). There, you can deposit a finished manuscript or an advanced interim version before the peer-reviewed publication has appeared. This makes the discovery available to the entire community, and documents one's priority. There is no peer review process, only a general check for plagiarism or inappropriate content. BioRxiv and similar services provide the opportunity for comments and discussions that can help improve the manuscript. In addition, by linking to other (social media) platforms, they allow you to make your work more widely known.

► Most peer-reviewed journals accept preprints in archives such as those mentioned above. However, you should make sure that this is the case for the journal to which you want to submit your article (and possibly for your second choice, in case you get a rejection…). Some journals even allow you to upload your manuscript directly from BioRxiv for official submission.

While the repositories described above mainly collect text documents, it is now part of the culture of open science to also make raw data or images available for download. Many journals require this when the manuscript is submitted, or at the latest when the article is accepted for publication. The traditional sentence "data will be made available on reasonable request" will probably not be sufficient in the future (Ascoli et al. 2017). But how and where can this be reasonably done? Guidelines and criteria developed in recent years by various professional societies and committees can help. As the field is developing rapidly, it is wise to keep abreast of the relevant and appropriate repositories. For example, the International Neuroinformatics Coordinating Facility (INCF; https://www.incf.org) and the EBRAINS initiative of the European Human Brain Project (https://ebrains.eu) provide specific information and direct access to databases. There are also specialized search engines for data or data repositories (e.g. https://datasetsearch.research.google.com).

An unresolved problem is the lack of standardization of data. Experiments are usually performed in lab-specific variants, which are difficult to compare with other labs; data are collected with different, often proprietary software and stored in the corresponding formats; metadata are hardly standardized at all; finally, different combinations of data types occur in complex experiments, whose combination can already pose a problem in one's own lab, let alone within the international community. Several organizations are actively working on standards and best practice examples (Teeters et al. 2015; Rübel et al. 2022). Unfortunately, these initiatives run in parallel and are not coordinated, so that generally accepted standards have emerged only in few specialized disciplines. These are usually fields with a fairly uniform data structures, such as functional imaging in humans. For electrophysiology, see the links in Table 7.1 and the organizations INCF, BNCN, and EBRAINS as mentioned above.

The demand for openness and transparency has led to the development of the FAIR Principles, which were originally formulated by a bottom-up initiative of

Table 7.1 List of some common repositories

Name	Web address	Type of data	Organization
Figshare	https://figshare.com	Very diverse: illustrations, raw data, manuscripts	Digital Science/ Springer Nature
Zenodo	https://www.zenodo.org	Various data sets from individual or group-coordinated projects	CERN, OpenAIRE, EU-Commission
GitHub	https://github.com	Mainly management of software and software versions, open source and interactive programming are possible	Microsoft
Neurodata Without Borders (NWB)	https://www.nwb.org	General standard for neurobiological data; references to numerous publicly accessible data sets	Various foundations and sponsors
Scientific Data	https://www.nature.com/sdata/	General electronic journal that publishes short descriptions of openly accessible data sets	Springer Nature
G-Node	https://www.g-node.org	Platform for sharing data, formats and analysis tools. Offers training, data sharing and analysis tools	INCF, BMBF, LMU München

numerous individuals and institutions and now enjoy wide international recognition (Wilkinson et al. 2016). Their implementation is being driven in Europe by the Go-Fair Organization (https://www.go-fair.org/), among others, but similar structures and initiatives can be found also in other countries, e.g. the USA. The FAIR Principles state that data should have four characteristics:

- **Findable:** It must be uniquely identified and tagged with rich, unambiguous metadata. It must be stored in a location (repository) where it is registered in such a way that it can be searched and found by humans and computers.
- **Accessible:** Data and metadata must be retrievable using a protocol/application that is freely accessible and may include a regulated procedure for authorizing users. Metadata should remain permanently accessible, even if the original data is no longer available.
- **Interoperable**: Data and metadata are stored in a form or language that is generally known and applicable in the scientific community. They refer to other data where necessary.
- **Reusable:** Data and metadata are well documented, licensed, and accompanied with all necessary information so that they can be used by others at any time. This includes adherence to domain-specific standards, in our case, the requirements and habits of the "patchers".

The whole thing may sound a bit abstract, but the intention is clear: data should be available to the scientific community for everyone to use on a permanent basis. Data sharing is increasingly becoming a mandatory part of public funding. It is clear that this requires a certain amount of effort and a mature infrastructure that, in addition to technical requirements, includes the definition of domain-specific standards, the establishment of interfaces for storage and retrieval of data, and, last but not least, the education of the community. Data science and data management are developing into new, demanding professions. For us as users, it is important to keep our eyes open and to use the resources that meet the FAIR Principles and are recommended by credible institutions. Here again, the above-mentioned organizations offer training, as do increasingly the universities themselves.

References

Abrams MB, Bjaalie JG, Das S et al (2021) A Standards Organization for Open and FAIR Neuroscience: the International Neuroinformatics Coordinating Facility. Neuroinform 20:25-36

Ascoli GA, Maraver P, Nanda S, Polavaram S, Armananzas R (2017) Win-win data sharing in neuroscience. Nat Methods 14:112–116

DFG: https://www.dfg.de/en/basics-topics/basics-and-principles-of-funding/research-data

FAIR Principles: https://www.go-fair.org/go-fair-initiative/

Klumpp M, Embray L, Heimburg F, Alves Dias AL, Simon J, Groh A, Draguhn A, Both M (2025) Syntalos: a software for precise synchronization of simultaneous multi-modal data acquisition and closed-loop interventions. Nat Commun 16:708

Sellers KK, Gilron R, Anso J et al (2021) Analysis-rcs-data: open-source toolbox for the ingestion, time-alignment, and visualization of sense and stimulation data from the medtronic summit rc+s system. Front Hum Neurosci 15:714256

Szell A, Martinez-Bellver S, Hegedus P, Hangya B (2020) Opeth: open source solution for real-time peri-event time histogram based on open ephys. Front Neuroinform 14:21

Teeters JL, Godfrey K, Young R, Dang C, Friedsam C, Wark B, Asari H, Peron S, Li N, Peyrache A et al (2015) Neurodata without borders: creating a common data format for neurophysiology. Neuron 88:629–634

Wilkinson MD, Dumontier M, Aalbersberg IJJ, Appleton G et al (2016) The FAIR Guiding Principles for scientific data management and stewardship. Scientific Data 3:160018

Index

F. Roth et al., *Mastering the Patch Clamp Technique: A Practical Guide*,
https://doi.org/10.1007/978-3-662-70745-6

MIX
Papier aus verantwortungsvollen Quellen
Paper from responsible sources
FSC® C105338

If you have any concerns about our products,
you can contact us on
ProductSafety@springernature.com

In case Publisher is established outside the EU,
the EU authorized representative is:
Springer Nature Customer Service Center GmbH
Europaplatz 3, 69115 Heidelberg, Germany

Printed by Libri Plureos GmbH
in Hamburg, Germany